OPTIMIZATION OF CAM MECHANISMS

SOLID MECHANICS AND ITS APPLICATIONS
Volume 9

Aims and Scope of the Series

The fundamental questions arising in mechanics are: *Why?*, *How?*, and *How much?* The aim of this series is to provide lucid accounts written by authoritative researchers giving vision and insight in answering these questions on the subject of mechanics as it relates to solids.

The scope of the series covers the entire spectrum of solid mechanics. Thus it includes the foundation of mechanics; variational formulations; computational mechanics; statics, kinematics and dynamics of rigid and elastic bodies; vibrations of solids and structures; dynamical systems and chaos; the theories of elasticity, plasticity and viscoelasticity; composite materials; rods, beams, shells and membranes; structural control and stability; soils, rocks and geomechanics; fracture; tribology; experimental mechanics; biomechanics and machine design.

The median level of presentation is the first year graduate student. Some texts are monographs defining the current state of the field; others are accessible to final year undergraduates; but essentially the emphasis is on readability and clarity.

For a list of related mechanics titles, see final pages.

Optimization of Cam Mechanisms

by

J. ANGELES
Robotic Mechanical Systems Laboratory,
McGill Research Centre for Intelligent Machines,
McGill University,
Montreal, Quebec, Canada

and

C. S. LÓPEZ-CAJÚN
Instituto Mexicano del Transporte,
San Fandila,
Queretaro, Mexico

KLUWER ACADEMIC PUBLISHERS
DORDRECHT / BOSTON / LONDON

Library of Congress Cataloging-in-Publication Data

Angeles, Jorge, 1943-
Optimization of cam mechanisms/ Jorge Angeles and Carlos S. López
-Cajún.
p. cm. -- (Solid mechanics and its applications ; 9)
Includes index.
ISBN 0-7923-1355-0 (alk. paper)
1. Cams--Design and construction. 2. CAD/CAM systems. I. López
-Cajún, Carlos S. II. Title. III. Series.
TJ206.A64 1991
621.8'38--dc20 91-23962

ISBN 0-7923-1355-0

Published by Kluwer Academic Publishers,
P.O. Box 17, 3300 AA Dordrecht, The Netherlands.

Kluwer Academic Publishers incorporates
the publishing programmes of
D. Reidel, Martinus Nijhoff, Dr W. Junk and MTP Press.

Sold and distributed in the U.S.A. and Canada
by Kluwer Academic Publishers,
101 Philip Drive, Norwell, MA 02061, U.S.A.

In all other countries, sold and distributed
by Kluwer Academic Publishers Group,
P.O. Box 322, 3300 AH Dordrecht, The Netherlands.

Printed on acid-free paper

Printed in the Netherlands

To the memory of my father
To my mother
Jorge Angeles

To the memory of my parents
Carlos S. López-Cajún

This is an account of how all was in suspense, all in calm,
in silence; all motionless, still, and the expanse of the sky was empty.

Popol Vuh, Part I, Chapter I
From (Goetz, 1954)

FOREWORD

Cam mechanisms are among the most reliable, economical and versatile means of producing periodic motion from a shaft rotating at uniform speed. Indeed, contrary to linkages, cam mechanisms can produce an intended motion program with an accuracy that is limited only by that of the machine used for cutting the cam. Moreover, the repeatability of a cam mechanism is basically zero, for it is perturbed only by wear, the effects of which can be predicted very reliably; in many cases wear can be virtually eliminated.

Technological developments in cam mechanisms, as in many other engineering fields, can be divided historically in two eras, namely, before the advent of the computer and afterwards. Before the computer became a commonplace aid in the design and manufacture of mechanical elements, cam drawing and cutting was more an art than a science. In fact, accuracy in both drafting and cutting of the cam profile was heavily dependent on draftsmen's and machinists' dexterity and technical skills. At present, the accuracy is limited only by the processor controlling the rendering of the cam profile and the media on which the rendering is provided, whether a plotter, a laser printer, a video monitor, or a machine tool. Processor performance is far better than that of these media.

On the other hand, developments in CAD and its supporting numerics and software have not been fully exploited in the design and manufacture of cam mechanisms.

its maxima and minima. Apart from this, we use the calculus concepts of slope and curvature of a curve described in polar coordinates. Where polar coordinates lead to cumbersome derivations we switch to Cartesian coordinates.

We recommend the use of cubic splines as a fundamental design tool, applicable not only to the synthesis of the follower-motion program, but also to that of the cam profile itself. For this reason, we give an account of periodic cubic splines, which are the simplest we have found applicable to cam design. However, for purposes of illustration, we include simple numerical examples based on two of the most common motion programs that, nevertheless, are quite suitable in the synthesis of the follower motion program, namely, the cycloidal and the 3-4-5 polynomial motion.

While we cover theoretical aspects of the optimum design process in full detail, we also look at technological aspects of the rendering of the cam profile with the aid of CAD/CAM software and hardware. In doing this, we do not assume any previous knowledge of this technology and thus explain all specialized terms and provide references for interested readers to become more acquainted with the subject. The principal aim of the book is to convince readers that an integration of the overall design process with the machining operations is possible with the aid of readily available software and hardware. We have achieved this integration at the laboratory level, using state-of-the-art computing hardware, namely UNIX color graphics workstations and an NC vertical machining center. In following this idea of computer integration, the book itself was typeset using INRSTEX, a TEX version running on the McRCIM network; TEXBINIM, a piece of software developed in-house for graphics was used for many figures; others were obtained with AutoCAD, a trademark of Autodesk Inc. The contents of the book is outlined in Section 1.5.

Parts of the book have been used in undergraduate courses on mechanisms, computer graphics, and computer-aided design. The book has been also used in its entirety in a graduate course on mechanism optimization. The book can be used in a regular undergraduate or graduate course, or for self study. The review problems in the Appendix are included to affirm the grasp of the subject matter and to suggest challenging problems that may lead to comprehensive term projects.

We are indebted to McRCIM for the use of their state-of-the-art network and supporting software and to all the individuals and institutions who supported our efforts in many ways. We especially want to thank: the McRCIM system administration staff for their help in coping with system failures; Max A. González-Palacios, a Ph. D. candidate at McGill's Mechanical Engineering Department, who developed the solid models and drawings of various cam mechanisms, with a high level of profesionalism; Marc Hasenohr, a visiting student from Ecole des Mines, in Paris, for the derivation of results in Subsection 2.4.3; M. Nishioka from Sankyo America, Inc., for granting us permission to reproduce the photographs included in Chapter 1; G. M. L. Gladwell, Kluwer's Series Editor for Solid Mechanics and its Applications, for proofreading the whole manuscript and, in so doing, for pointing out technical flaws that otherwise would have remained undetected; Leonor R. Delgado, for her help in correcting English and style; Helga Symington, for her patience and skills in taking the photographs of Max's solid models and prototypes.

Early methods of synthesis of follower-motion programs were based on a very limited set of primitive functions leading to *uniform, parabolic, harmonic, cycloidal* and, more recently, *polynomial motions.* Of these, the most versatile motion functions are the latter, for they offer the highest number of parameters to choose from in producing a desired motion. Morever, finding the most suitable values of these parameters requires the solution of linear equations, which is what computers are meant for, in the first place. However, polynomial motions have an inherent drawback that tends to be overlooked by the designers, namely, the numerical stability of the associated equations, which deteriorates very rapidly as the degree of the polynomial increases. Therefore, alternative solutions must be explored.

Among the most suitable means of producing complex shapes with prescribed local or global properties, *splines* appear as the most viable alternative, for they lead to inherently stable numerical procedures. Splines have only recently and rather slowly found their way into the realm of cam design. The most popular splines in this context are *B-splines.* The authors have experimented with cubic splines for the last ten years and have found that they are very suitable and *user-friendly* elements, not only for the accurate design of motion programs, but also for the optimization of these and the cam profiles. This book is a summary of the research conducted by the authors in this period, at various places. The first author became interested in splines while on a sabbatical leave at the Machine-Tool Laboratory of the Aachen Institute of Technology—*Werkzeugsmaschinenlaboratorium der Rheinisch-Westfälischen Technischen Hochschule Aachen*—in the Federal Republic of Germany, under an Alexander von Humboldt Fellowship. This work was later continued at the CAD Laboratory of the Faculty of Engineering of the National Autonomous University of Mexico (UNAM), where the second author joined the first in this endeavor. Since 1984, this work has been conducted at the Robotic Mechanical Systems Laboratory of the McGill Research Centre for Intelligent Machines (McRCIM), McGill University. An *Actions structurantes* Fellowship, granted to the second author by the Government of Quebec, allowed him to continue this work at McRCIM for two years.

Our aim in writing this book is to present the design of cam mechanisms as a unified optimization process. The approach is original, systematic, unified, and simple. We depart from the standard practice of cam design in that we focus on optimization from the very beginning. In doing this, we have devised methods that are especially applicable to the types of cam mechanisms discussed here. While we present the theoretical results with all detail and rigor, we derive practical design data in the form of charts or figures of merit, that can be readily used in any design studio. We also depart from the common practice of deriving the equations of the cam profile using envelope theory. Although this theory is not a very advanced mathematical tool, it is not necessarily contained in a regular engineering curriculum, and hence we prefer to dispense with it. Thus, we derive the coordinates of the cam profile using simple geometry and trigonometric identities. Moreover, we derive all our results on cam optimization by resorting only to basics, namely, conditions upon which a smooth function, like the pressure angle in roller-follower cam mechanisms, attains

Last, but not least, we thank: the Alexander von Humboldt Foundation; UNAM; the *Fonds pour la formation de chercheurs et l'aide à la recherche* of Quebec; the *Natural Sciences and Engineering Research Council* of Canada; the *Instituto Mexicano del Transporte*; and the Department of Mechanical Engineering of McGill University.

Montreal, May 1991

Jorge Angeles and Carlos S. López-Cajún

CONTENTS

Chapter 1 INTRODUCTION

1.1 Preliminary Concepts

A cam mechanism is a mechanical system consisting of three basic components: a driving element, called the *cam*; a driven element, termed the *follower*; and a fixed frame. Sometimes, an intermediate element is introduced between the cam and the follower with the purpose of improving the mechanism performance. This element is called the *roller* because function is to produce a pure-rolling relative motion between the cam and the follower. The purpose of a cam mechanism is the transmission of power or information. In applications concerning power transmission, the main good to be transmitted is force or torque; in applications of information transmission, the main good transmitted takes the form of motion signals. Most modern applications of cam mechanisms, to be described shortly, are of the former type. Cam mechanisms used for information transmission were traditionally found in measuring instruments. With the advent of modern microprocessor-based hardware, this type of application is becoming less common. Nevertheless, cam mechanisms are still used in a wide spectrum of applications, especially in automatic machines and instruments, textile machinery, computers, printing presses, food-processing equipment, internal combustion engines, control systems, and photographic equipment (Prenzel, 1989).

In the design of cam mechanisms, the engineer performs several activities, namely, task definition, synthesis, analysis, optimization, and dynamic simulation. These tasks do not always follow this order. In fact, some loops may appear in the foregoing tasks, such as those illustrated in Fig. 1.1.1. In that figure, the decision blocks with the "OK?" symbol stand for performance evaluation, which may certainly involve a considerable amount of analysis. In the first such block, the evaluation may most likely be limited to kinematic and elastostatic analyses. In the second decision block, the evaluation will involve the dynamic analysis of the entire mechanical system, of which the cam-follower mechanism is only a component. The main issues in dynamic evaluation involve vibration response and dynamic stability. Both elastostatic and dynamic analysis will surely require the introduction of finite-element methods.

A more detailed description of the various blocks appearing in Fig. 1.1.1 follows: *task definition* refers to translating the engineering need motivating the design

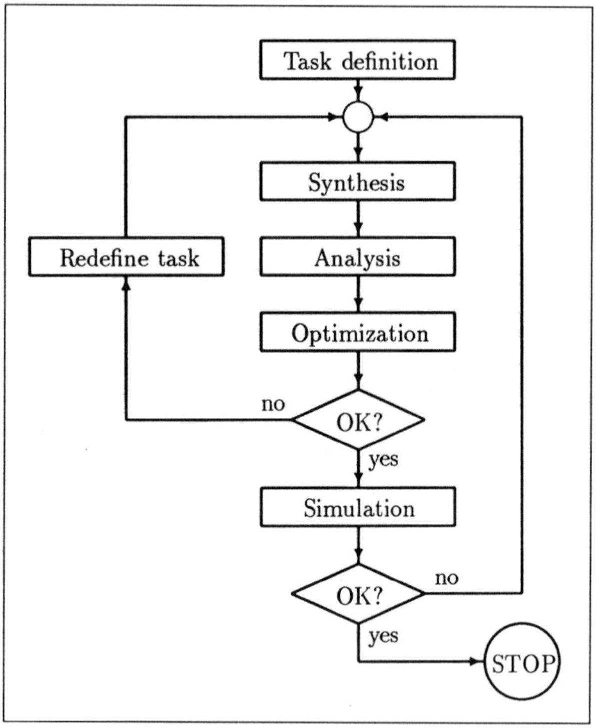

Fig. 1.1.1. Engineering activities related to cam design.

at hand into a *follower displacement program*. This is a *displacement history*, i.e., a function $f(t)$ of time that can represent a translational displacement in the case of a translating follower or an angular displacement in the case of an oscillating follower. Moreover, this function can either be given continuously over the time interval under consideration, $[0, T]$, or discretely. Furthermore, as is usually the case, the displacement program can be given only as a sequence of stages, namely, dwell during aT, rise during bT, dwell during cT, and return during dT, where a, b, c, and d are positive fractions that add up to 1, without specifying how rise and return take place. The only specification may be that the dwell stages be *blended* in a smooth fashion, i.e., with zero velocity and acceleration at transition instants.

This is the most common type of task definition.

The most basic of the activities outlined in Fig. 1.1.1 are essentially of a kinematic nature. These are, in fact, the core of the cam design process, and they are the focus of this book. The basic assumption here is that both the cam and the follower are rigid bodies. Additional elements, such as rollers in the case of roller followers, are assumed to be rigid as well.

Given a task definition in the form of a follower-displacement program $f(t)$, the *synthesis* stage consists of producing a cam surface, in general, with the shape required to produce the desired motion. In the case of disk cams, the said surface is cylindrical, and hence, is fully determined by its generating curve, which is known as the *cam profile*. In a system of polar coordinates fixed to the cam disk, this profile is given by a functional relation $\rho(\theta)$, where ρ and θ are the polar coordinates of the cam profile. Usually, $\rho(\theta)$ is not derived directly but in parametric form as $\rho = \rho(\psi)$, $\theta = \theta(\psi)$ or, equivalently, via the Cartesian coordinates of a point of the cam profile, i.e., $x = x(\psi)$, $y = y(\psi)$, where ψ denotes the angle of rotation of the cam. In the case of cylindrical cams, the cam surface takes on a more complex form, but it is still a ruled surface, which is fully determined by the trajectory of the generating line that describes a surface known as the *cylindroid*, to be discussed in more detail in Chapter 6.

The next step in the design process is *analysis*, which consists of, given a cam-follower mechanism, i.e., the cam surface and the follower geometry, determine the follower displacement program. This is a step that is performed mainly for purposes of design evaluation. Alternatively, one may want to determine the type of motion produced by a cam profile that may be available either through a database or as a machined part. In the latter instance, analysis of the cam mechanism requires some pre-processing, namely, digitizing the cam profile. This can be done in several ways, depending on the type of hardware at hand. If it is possible to obtain an accurate drawing of the cam profile, then the cam profile can be represented in a database using a digitizer, as shown in Fig. 1.1.2. A sample of point coordinates is recorded and stored in memory for further processing. Once the cam profile is stored in a database, the cam mechanism analysis is straightforward, as indicated in Chapter 7.

Cam *optimization* refers to determining the cam surface producing a given follower displacement program that minimizes a certain *cost* or maximizes a certain *profit*. A cost is any item that one is interested in keeping low for purposes of economy, such as material volume, inertia of moving parts, stress levels, and so on, whereas a profit is any item that enhances the operation of the mechanism, e.g., force transmission, reliability, life span, and the like. In a general context, particularly at the stage of problem definition, one does not know if the problem at hand will lead either to cost minimization or profit maximization. In this case, we speak, in general, of optimization of an *objective function*.

Most frequently, optimization cannot be performed by freely choosing the *design variables* of the problem. In the case of cam optimization, these variables determine the shape of the cam surface. In fact, the designer usually must observe a set of functionality and feasibility constraints. These involve various items such as limited resources, force, energy, and space available. In this case, we speak of *constrained optimization*, to distinguish this problem from the simpler one that is known as *unconstrained optimization*.

Fig. 1.1.2. Digitalization of a cam profile.

In the realm of cam design, optimization can lead to very complex problems requiring the use of sophisticated numerical methods that fall beyond the scope of this book, which is kinematic design. Hence, items, such as structural optimization and design optimization for manufacturability and assembly, will not be studied. Given this scope, the optimization methods that we will apply are rather straightforward. They involve, in general, the minimization or the maximization of a real-valued function $z(x)$ of a real variable x. Moreover, we will sometimes impose the condition that, at the optimum, z attain a prescribed value z_0. The *necessary* condition for a maximum or a minimum is that the derivative of z with with respect to x vanishes. Thus, the equations that will allow us to find the optimum values of our design variables are

$$z'(x_0) = 0, \quad z(x_0) = z_0 \tag{1.1.1}$$

where x_0 is the particular value of x at which the optimum is attained. Note that the above condition is necessary, but not sufficient, for all that this condition guarantees is the presence of a stationary point, which can be a maximum, a minimum or a saddle point. We should recall that minima occur for a positive second derivative of z with respect to x, maxima for a negative second derivative. Saddle points occur when the second derivative vanishes. However, the latter conditions, although sufficient, are not necessary. For example, the function x^{2k}, for any integer value of k, has one minimum at $x = 0$, where all its derivatives vanish. In the problems we will study, the nature of the stationary point found using the first-order necessary condition (1.1.1) will be derived by inspection.

Traditional methods of cam design mainly involve analysis and synthesis, touching only marginally upon the subject of optimization. The availability of powerful computing and graphics software and hardware makes it possible to pursue the optimization of cam profiles at the outset, as demonstrated in this book. Moreover, we show that, for the types of cams considered here, simple one-dimensional optimization procedures can be successfully applied. Alternatively, some cam profiles can be optimized by verifying a *normality condition*, which involves finding the solution of a single-variable nonlinear equation. In this case, the said solution can be found visually with the aid of suitable graphics displays and the necessary supporting software. Alternatively, it can be found using an iterative numerical method.

Once the cam profile has been obtained and its inertia characteristics determined, *simulation* comes into the picture. This stage is intended for verifying the dynamic behavior of the cam-follower system, i.e., *what is the response of the system when it is driven at various speeds or when it is subjected to the action of different loads?*

The coupling between cam and follower, or, alternatively, between cam and roller, takes place along a line or a point if we look only at disk cams and their projection onto the plane of motion. Such a coupling determines what is called a *higher kinematic pair*, as discussed in Section 1.3. A characteristic of higher pairs is that the relative motion between coupled elements is a combination of rolling and sliding of one rigid body with respect to the other. Moreover, unlike linkages, this type of mechanism allows us to produce an output $f(\psi)$, in the form of a function of the output ψ, with *total precision*. That is, whereas the design parameters of linkages form a finite set—the link lengths and offsets, as well as the angles between consecutive axes (Suh and Radcliffe, 1978; Angeles, 1982)—, the design parameters of cams form an infinite set or continuum. This continuum is given by the whole set of points constituting the cam surface or the cam profile in the case of disk cams.

In considering the representation of the follower displacement program, the traditional approach has been based on a limited set of functions giving rise to parabolic, harmonic, cycloidal, trapezoidal, polynomial motions, and combinations thereof (Rothbart, 1956). Only recently was a totally new approach introduced using spline functions (Hitoshi, 1976; Nó Sánchez and García de Jalón, 1980; Angeles, 1983a).

As for the profile synthesis, the classical method based on graphical techniques has been abandoned in favor of numerical methods which are more appropriate to an age of computers and numerically-controlled machine tools (Grant and Soni, 1979;

Chen, 1977).

In this book, the approach to the synthesis of both the displacement program and the cam profile will be based on both computer graphics and numerical methods. In the sections below, first a brief historical review is included; then, the classification of cam mechanisms is presented, and is followed by a discussion of general design considerations.

1.2 Historical Notes

Cam mechanisms have their origin in the *wedge*, one of the *simple mechanisms* of the ancient times. In fact, Müller (1987) claims that the origin of cam mechanisms dates from the Paleolithic age, and hence, cam mechanisms may have appeared about 10, 000 years ago. Hero of Alexandria is recognized as having used a primitive cam mechanism as a leveling instrument in the first century B. C. (Müller, 1987; Dudiţa et al., 1989). We find in the same references that cam mechanisms were used in the Middle Ages to drive bellows and foundry hammers. However, it is the genius of Leonardo da Vinci that lies at the origin of the modern design of cam mechanisms (Müller and Mauersberger, 1989; Dudiţa et al., 1989). More recently, cam mechanisms were developed by the clasical kinematicians of the XIXth century, as described with particular detail by Dudiţa et al. in the same reference. Below we give a short account of these authors' historical notes.

The introduction of cam design, according to the aforementioned reference, is a legacy of the French Revolution, and is particularly attributed to the French geometer L. Monge, who is credited with the creation of descriptive geometry. A method for the synthesis of the profile of a cam mechanism is reported by the French J. Leupold in 1724, whereas Lanz and Bétancourt, professors at the *Ecole Polytechnique*, reported a method of cam drawing in their textbook *Essai sur la composition des machines*, written in 1808 and published in 1811. They attributed that method to Deparcieux, who first presented it to the Academy of Sciences in 1747. In 1811, as well, Hachette published a handbook under the title *Traité élémentaire des machines*, describing various methods of cam-profile drawing. Another contributor to the design of cam mechanisms is J. Borgnis, professor at the University of Pavia, who published his *Traité complet de mécanique* in Paris in 1818, including cylindrical cams. One more design of a cylindrical cam was reported by A. Schubert, professor of mechanisms at the Technical University of Dresden, in 1829, while J. V. Poncelet is to be credited with the introduction of the *pressure angle*, in 1826, in his lectures on mechanics. A breakthrough in the development of cam mechanisms is the work of R. Willis, professor at Cambridge University, who completed in 1841 the first draft of his *Principles of Mechanism*, first published in 1875 in book form. In this book, Willis introduced a classification of cam mechanisms and proposes various designs of mechanisms for the production of harmonic motion, among others the Scotch yoke, not a cam mechanism but rather a lower-pair mechanism (Hartenberg and Denavit, 1964). Willis also included various designs of spatial cams in his book. By the end of the XIXth century, cam mechanisms were taught regularly at universities in Britain, France, Germany, and Russia. The study of cams was enhanced in 1900 with the appearance of *Theoretische Kinematik* by F. Reuleaux, professor at Braunschweig.

The design of cam mechanisms in the first half of the XXth century was based on the geometrical methods devised in the XVIIIth and XIXth centuries that relied on drafting instruments, and hence, were highly dependent on draftsmen's dexterity. Likewise, the manufacture of cams was done manually, using saws for a rough cutting of the metal plate, followed by a fine finishing with the use of a file. Obviously, the accuracy of cam manufacture was totally dependent on manual skills as well. Manual methods of drafting and cutting cams were not only inaccurate, but also time consuming. These methods changed with the advent of the microprocessor in the sixties that made possible the computer-controlled motion of both plotters and machine tools, thereby giving rise to CAD/CAM technology. This technology brought about a tremendous increase in the accuracy of cam-profile drawing and cutting. In fact, with the introduction of the cathode-ray tube as a means of displaying a computer-generated contour, the need for a cam-profile blueprint was virtually eliminated. Nowadays it is possible to produce a cam disk without ever producing a drawing of the cam contour. Moreover, since the process of cam synthesis can be totally automated, the designer is free from the tedious calculations required for design verification.

With all the advances in CAD/CAM hardware and software, however, cam design has remained essentially at the stage of development attained at the turn of the century. In fact, in spite of the availability of fast processors for number crunching and graphical display, few attempts aimed at the *optimization* of cam mechanisms have been reported. Optimization, to be discussed in greater detail later, allows the production of not only a *feasible* design, but also of the *best* design, provided that a measure of design *goodness* has been previously defined. This book is about the *design optimization* of cam mechanisms. Before we proceed further we describe some terminology which will be used throughout this book.

1.3 Nomenclature and Classification of Cam Mechanisms

In speaking of cam mechanisms, we must first agree on the concept of a mechanism. The *Concise Oxford Dictionary* defines a mechanism as a system of mutually adapted parts, whereas the *Random House Dictionary* states that a mechanism is an assembly of moving parts performing a complete motion. In the two foregoing concepts, the basic component of a mechanism appears to be a set of assembled parts. In kinematics, these parts are understood as rigid bodies, which are defined below. If *any* two points of a body undergoing an arbitrary motion remain equidistant, the body is said to be rigid. In kinematics, a rigid body is called a *link*, the assembly of a set of links constituting a *kinematic chain*. Links in a kinematic chain are coupled pairwise, and hence, each coupling constitutes a *kinematic pair*. Moreover, every coupling can take on two different forms, the first of which is along a surface and the second is either along a line or at a point. In the former case, a *lower* kinematic pair exists; in the latter, we have a *higher* kinematic pair. Lower kinematic pairs can be of six types, namely, revolute, prismatic, screw, cylindrical, spherical, and planar (Angeles, 1988). In fact, two basic kinematic pairs, the revolute and the prismatic pair, can be combined to form the remaining four. In cam mechanisms, the coupling between cam and follower, or cam and roller, takes place via a higher kinematic pair, whereas other couplings are of the lower type.

In these mechanisms, the cam usually moves with uniform speed, whether under pure rotation or pure translation with respect to the frame. On the contrary, the follower moves, either under pure rotation or pure translation with respect to the frame, according to a prescribed *displacement program* that includes four basic phases, namely, the lower dwell, the rise, the upper dwell, and the return. Thus, whereas the cam undergoes a uniform motion, the follower undergoes a *periodic* motion. Henceforth, the follower displacement program, whether translational or rotational, will be assumed to be composed of a constant plus a positive-semidefinite function $\sigma(\psi)$, where ψ denotes the variable defining the motion of the cam, and hence, $\dot{\psi}$ is constant. Cams to be studied in this book are rotating, ψ thus denoting angular displacement. Moreover, the positive-semidefiniteness of $\sigma(\psi)$ implies that this function is non-negative throughout its range of definition, that is, for $\psi \in [0, 2\pi]$, and does not vanish throughout the entire interval. Furthermore, $\sigma(\psi)$ attains its minimum value of 0 at the lower dwell, whereas, at the upper dwell, it attains its maximum value h. For translating followers, the displacement program will be denoted by $s(\psi)$ whereas for rotational followers it will be denoted by $\phi(\psi)$. Moreover, because of the uniform velocity motion of the cam, the foregoing motions can be alternatively represented as $\sigma(t)$, $s(t)$, or $\phi(t)$.

In a cam-follower mechanism, both the cam and the follower can be either rotating or translating. In the first case, an axis of motion, namely, an axis of rotation, is clearly defined. In the second, a *direction* of translation is defined, but no axis can be defined, for all points of the element under consideration, whether cam or follower, undergo the same displacement. Clearly, a rotational motion exists when the coupling between the moving element and the frame is of the *revolute* type; the motion is translation when the coupling is prismatic. For brevity, however, reference will be made to the axis of the cam or follower, regardless of whether the associated motion is translational or rotational.

Cam mechanisms are classified according to various criteria, as listed below:

- the relative layout of the axes of the cam and the follower
- the output type of motion
- the follower shape and layout

- the type of motion of the cam disk
- the shape of the cam disk
- the type of contact between cam and follower
- the number of working cycles of the follower

1.3.1 Classification According to the Layout of the Cam and Follower Axes

This classification, applicable when both cam and follower undergo rotational motion (Dittrich, 1970; Volmer, 1976; Müller and Kassamanian, 1979), is given below:

- *planar*, shown in Fig. 1.3.1. The cam and the follower axes of motion are parallel to each other. The literature on this type of cams is very rich, as the extensive survey reported in the References suggests.
- *spherical*, shown in Fig. 1.3.2. In this case, the cam and the follower axes of motion are concurrent. The literature on this subject is rather scarce, a comprehensive discussion being reported by Dittrich (1966).
- *spatial*, shown in Fig. 1.3.3. The axes of motion are skew.

1.3.2 Classification According to the Output Type of Motion

This classification is based on the type of motion that the follower can undergo, namely,

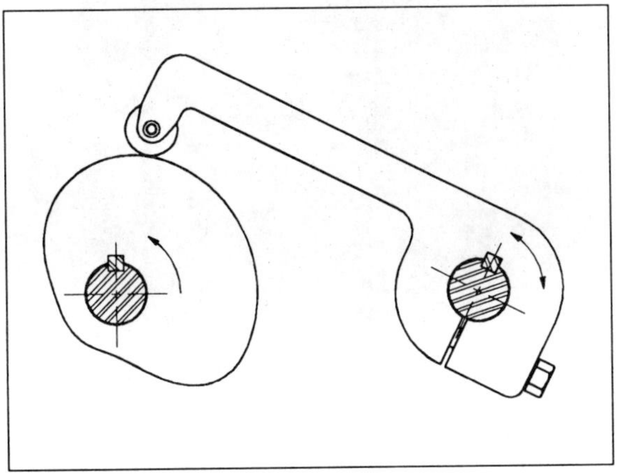

Fig. 1.3.1. Planar cam-follower mechanism.

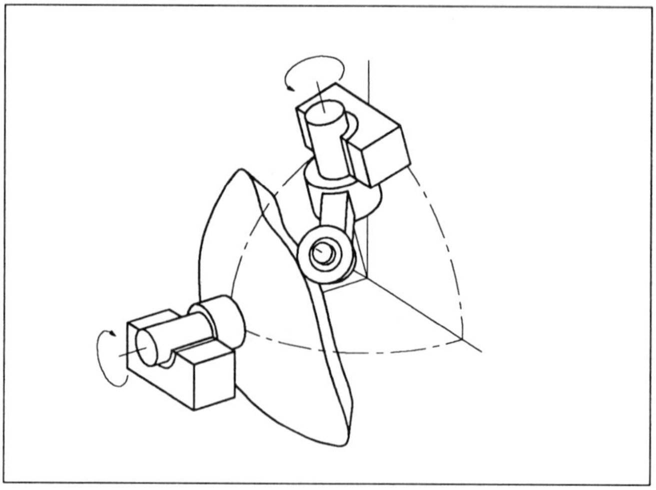

Fig. 1.3.2. Spherical cam-follower mechanism.

Fig. 1.3.3. Spatial cam-follower mechanism.
(Courtesy of Sankyo America, Inc.)

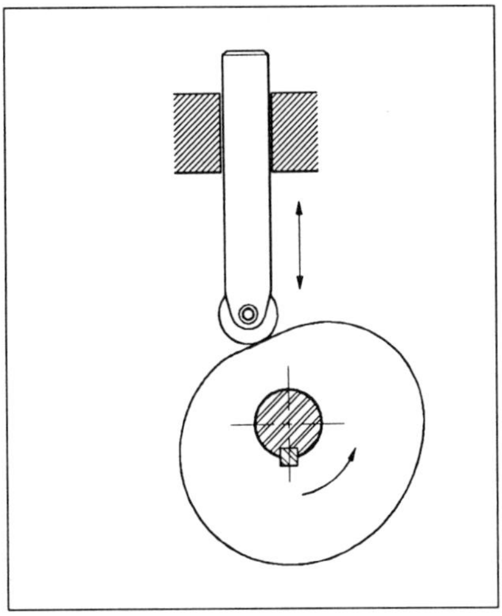

Fig. 1.3.4. Disk cam with translating follower.

- *translating* followers, as shown in Fig. 1.3.4. The follower undergoes pure translation
- *oscillating* followers, when the follower undergoes pure rotation, as shown in Fig. 1.3.1.

1.3.3 Classification of Followers

Followers can be classified according to:

- their physical shape at the coupling interface. These shapes are: *knife-edge*, *roller*, and *flat-face* followers, Figs. 1.3.5*a*, *b*, and *c*, respectively. Followers of the first type, however, are practically unsuitable for heavy-duty applications. Nevertheless, they are kinematically equivalent to roller followers, and hence their inclusion is justified, as will be seen in Chapter 5.
- their location with respect to the cam axis of rotation. This gives rise to *radial* and *offset* followers, as shown in Figs. 1.3.6*a* and *b*, respectively. When the path of the contact point of a knife-edge follower or of the roller center of a roller follower intersects the cam axis of rotation, the first type results; otherwise, it is the second type. In the latter, the offset causes a reduction of the thrust present in a knife-edge—or roller-follower. Thus, one can have more compact mechanisms for the same purpose if some offset is introduced, as discussed in

Chapter 5. It is pointed out that the offset concept for a flat-face follower is kinematically irrelevant and need not be considered.

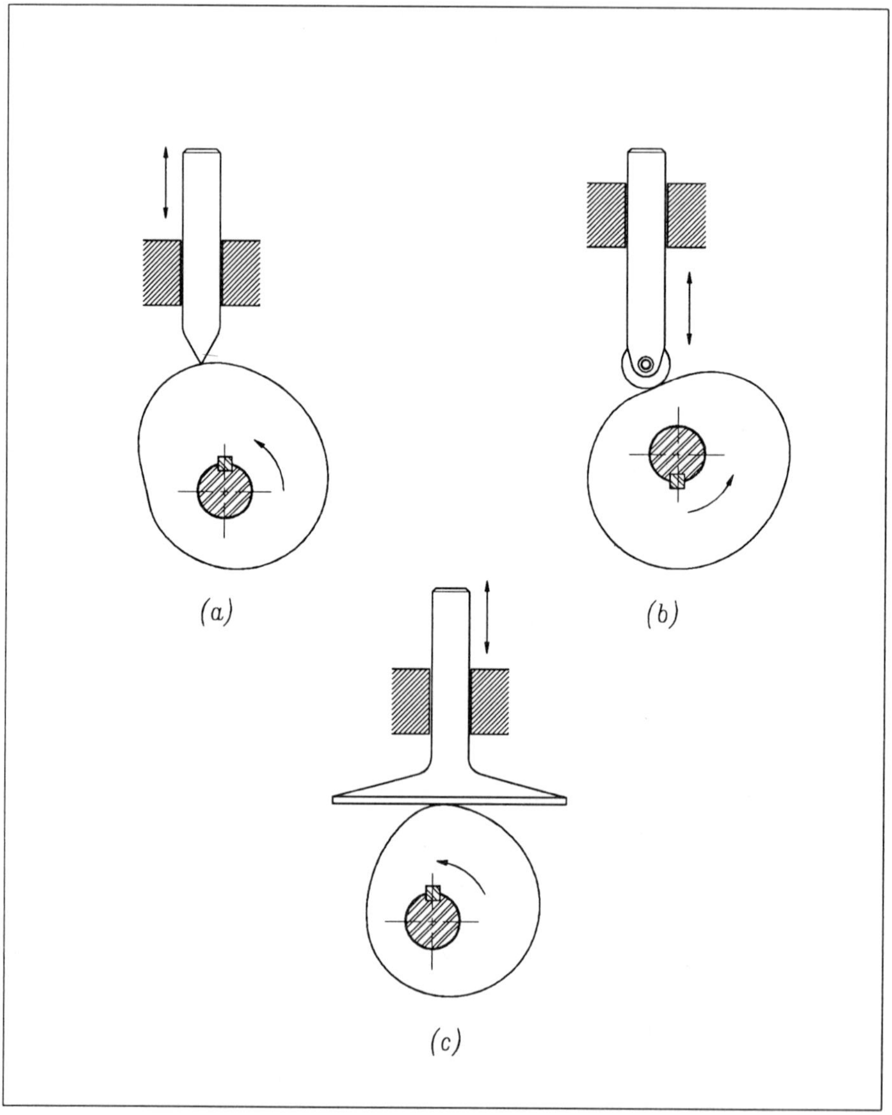

Fig. 1.3.5. Disk cams: *a*) Knife-edge follower; *b*) Roller-follower; and *c*) Flat-face follower.

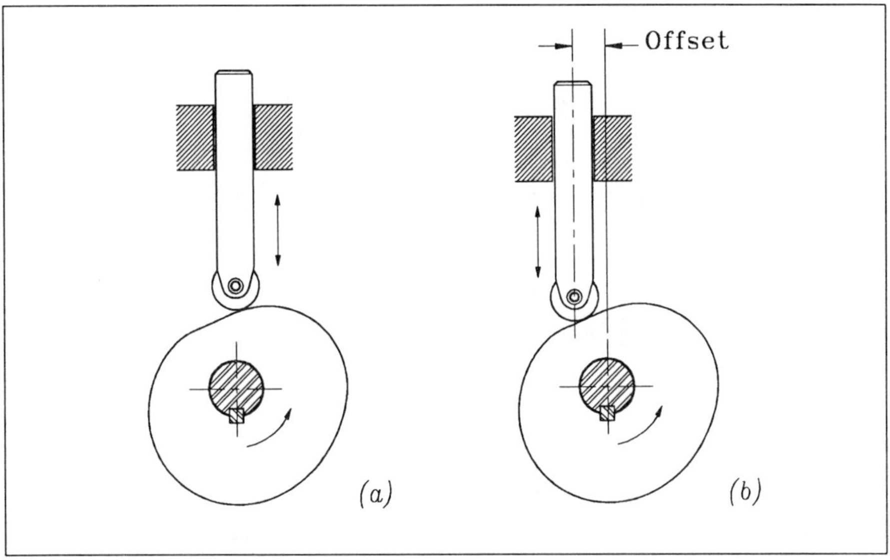

Fig. 1.3.6. Disk cams: *a*) Radial follower; and *b*) Offset follower.

1.3.4 Classification of Cams According to Their Type of Motion

Cam disks can be classified according to their type of motion. Thus, there are

- *translating* cams, as the one shown in Fig. 1.3.7.
- *rotating* cams. These are the most common. They are usually designed assuming that their angular velocity is uniform, Fig. 1.3.8.

1.3.5 Classification of Cams According to Their Physical Shape

According to the physical shape of the cam disk, there are

- *disk or plate* cams. These are the most widespread due to their simplicity and compactness. In these cases, the axis of rotation of the cam is perpendicular to the plane of the follower motion (Figs. 1.3.4–1.3.6)
- *wedge* cams. This type of cams is shown in Fig. 1.3.9. Their design is simple, but they have very limited applications. These are basically 3-link lower-pair closed kinematic chains.

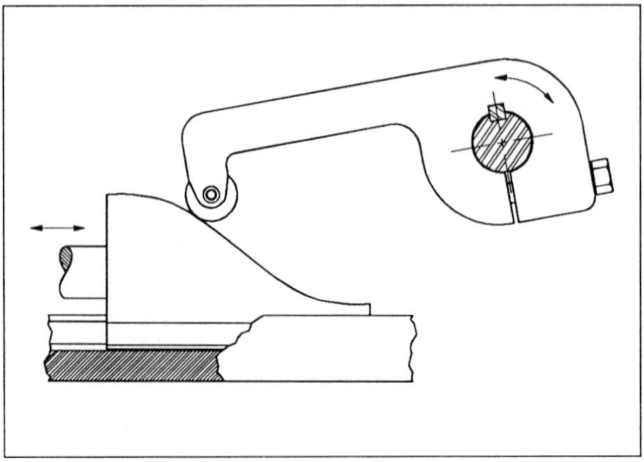

Fig. 1.3.7. Translating cam.

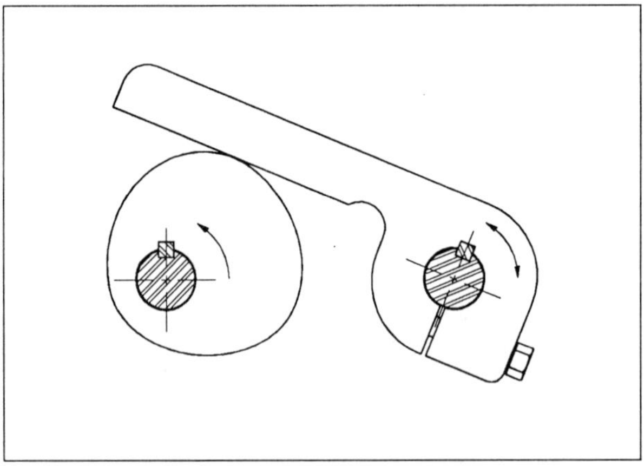

Fig. 1.3.8. Oscillating cam.

- *cylindrical* and *conical* cams. In both cases, the follower is constrained to move in a groove cut on the surface. If the surface is a cylinder the first type results; if it is a cone, the second type occurs (Figs. 1.3.10*a* and *b*, respectively).

- *spiral* cams. One such cam is shown in Fig. 1.3.11. The disadvantage of this type is that the cam movement has to be reversed at each cycle.

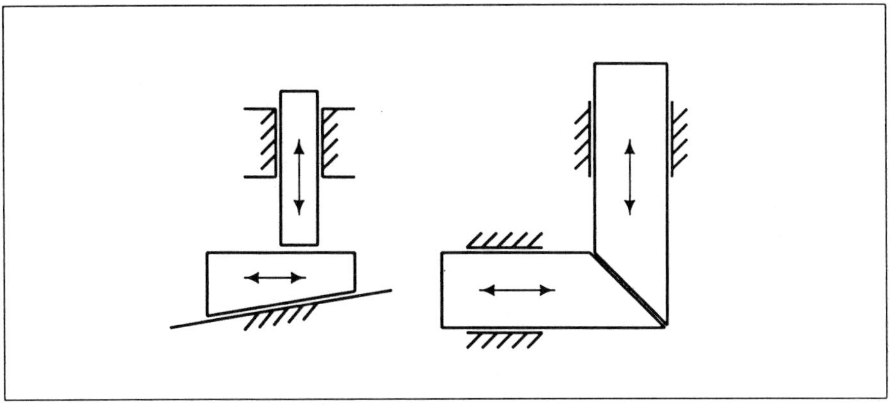

Fig. 1.3.9. Wedge cams.

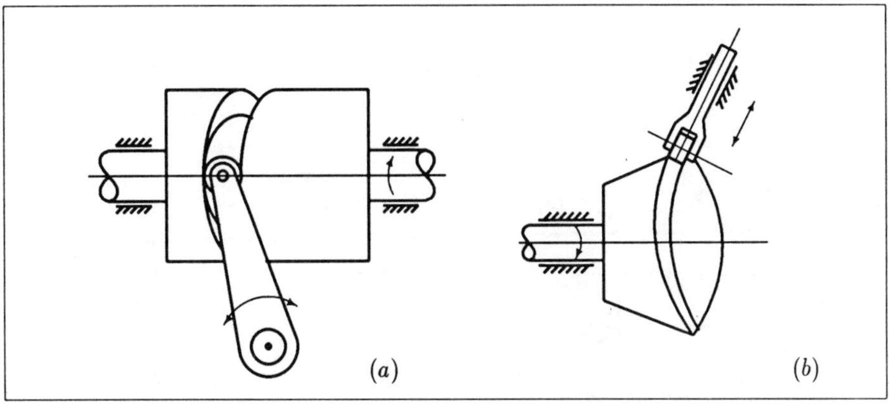

Fig. 1.3.10. *a*) Cylindrical and *b*) conical cams.

- *spherical* cams, such as in Figs. 1.3.2. The axis of rotation of the cam intersects the follower-motion axis.

- *globoidal* cams. This type can be either concave, Fig. 1.3.12*a*, or convex, Fig. 1.3.12*b*. The cam, which rotates about its axis of symmetry, has a circumferential contour cut on its surface. This type of cams is used for indexing mechanisms—see subsection below as well as (Rees Jones, 1978; Backhouse and Rees Jones, 1989).

- *conjugate*. In this type of cams, contact takes place at two points, which allows the transmission of moment or force in two directions, thereby eliminating the need for springs. A cam of this type is illustrated in Fig. 1.3.13. From this figure it is clear that the cam is composed of two rigidly attached plates with the same shape, one rotated or reflected with respect to the other. The same effect can be obtained with positive-motion cams, as described in the subsection below. However, unlike positive-motion cams, conjugate cams allow for pure rolling of all surfaces in contact, and hence, can withstand higher loads with less friction losses.

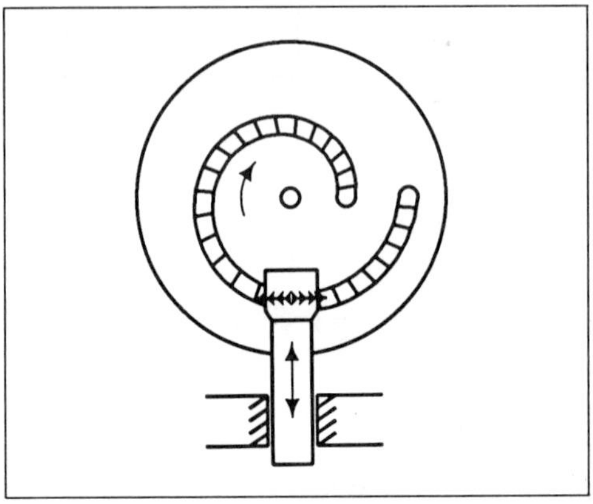

Fig. 1.3.11. Spiral cam.

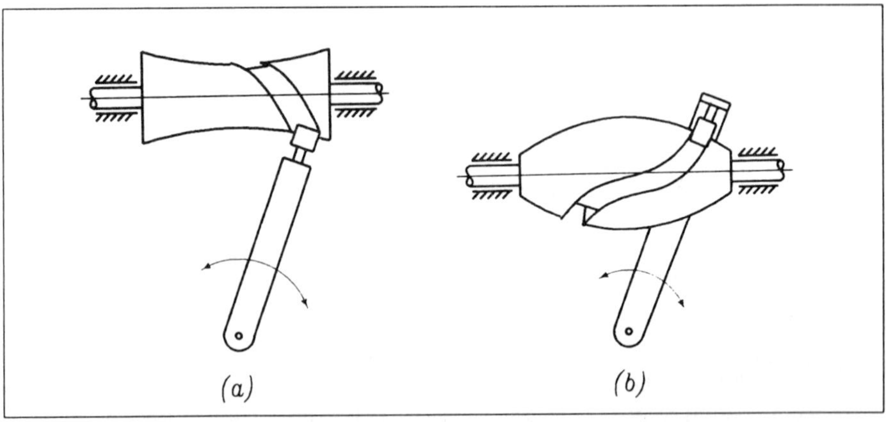

Fig. 1.3.12. *a*) Concave and *b*) convex globoidal cams.

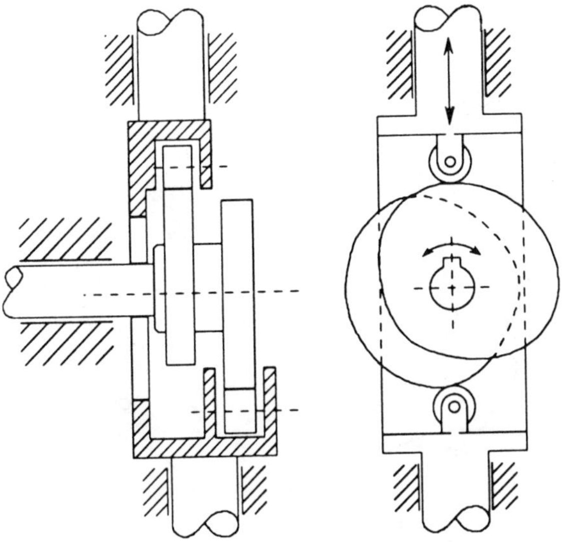

Fig. 1.3.13. Conjugate cam mechanism.

1.3.6 Classification According to the Type of Contact Between Cam and Follower

According to this classification, contact between cam and follower can be either simple or positive. Simple contact occurs when the cam can exert a one-directional force upon the follower, such as in the cases depicted in Figs. 1.3.1, 1.3.2, and 1.3.4–1.3.8. Clearly, in this case an external force is needed to keep contact at the return phase. This is usually achieved by means of springs, as shown in Fig. 1.3.14.

Positive contact exists between cam and follower when the cam can exert a bidirectional force on the follower, thereby eliminating the need for springs. Examples of this type of cam-follower mechanism appear in Figs. 1.3.15*a* and *b*. When positive contact exists, one speaks of a positive-motion cam. Note that positive cams entail a serious friction problem for, as illustrated in the foregoing figure, although pure rolling can be achieved between the roller and one of the cam tracks, say, the inner track, the relative motion between the roller and the other track is bound to produce sliding, unless a clearance is provided. The introduction of clearance, however, brings about problems of backlash.

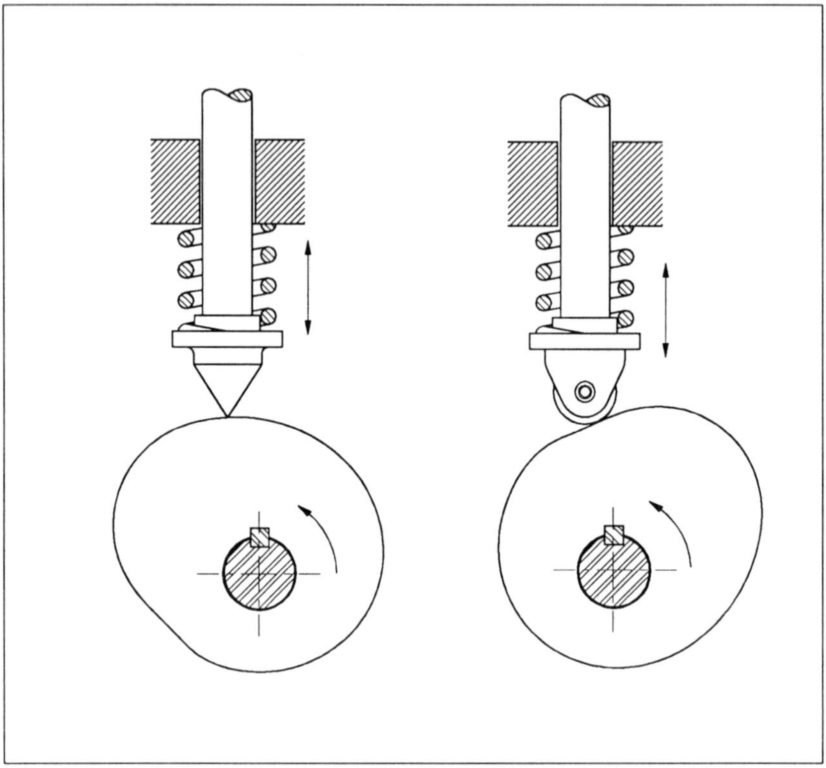

Fig. 1.3.14. Spring-loaded cam mechanisms.

1.3.7 Classification According to the Number of Working Cycles of the Follower

Finally, a follower can return to its lower dwell after the upper dwell is completed. In this case, one can speak of a *monocyclical* cam-follower mechanism. If, on the contrary, the follower returns to its lower dwell after having reached several intermediate upper dwells, the mechanism is of the *indexing* type. Indexing cam-follower mechanisms (Makino, 1976) have many applications in industry, in tasks requiring, for example, the performance of various similar operations onto a certain workpiece, such as boring holes of various diameters and depths. Substantial increases in productivity are achieved by performing as many operations on the workpiece as possible with the same machine. This is readily achieved using, for example, a vertical machining center supplied with an indexing mechanism presenting the various types of boring tools to the same workpiece; the latter might require to be properly positioned with respect to the tool for each operation. This positioning is achieved, in turn, by means of a computer-controlled two- or three-axis table. Indexing cam mechanisms can be of the planar, spherical, or spatial type. Planar and spatial mechanisms are illustrated in Figs. 1.3.16 and 1.3.17, respectively. In addition, Fig. 1.3.18 shows a photograph of an indexing cam mechanism with six indexing steps.

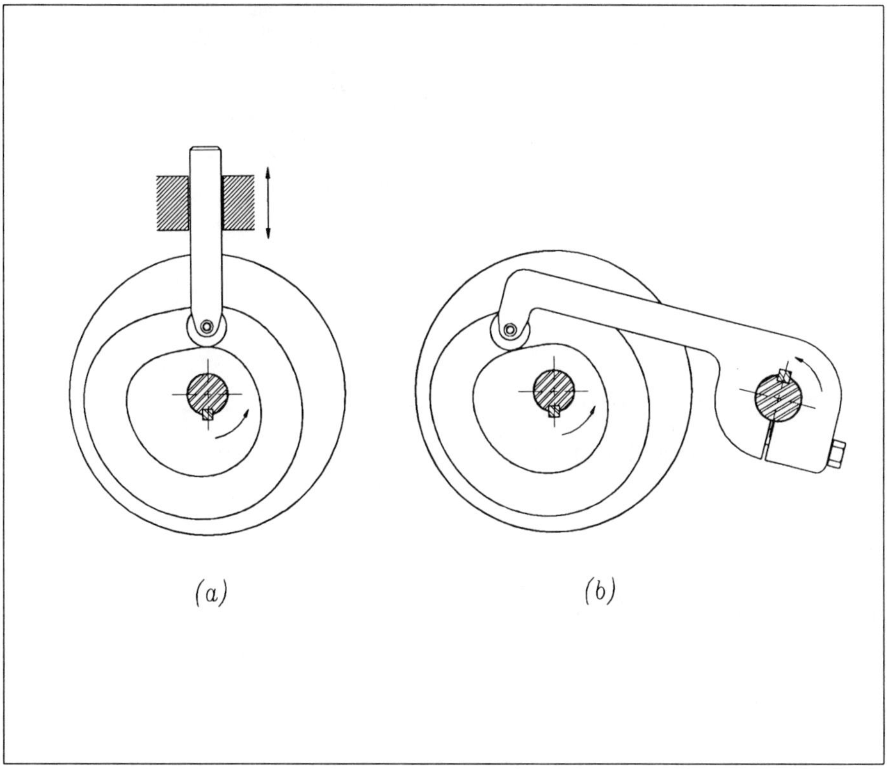

Fig. 1.3.15. Positive cam mechanisms: *a*) translational follower; *b*) oscillating follower.

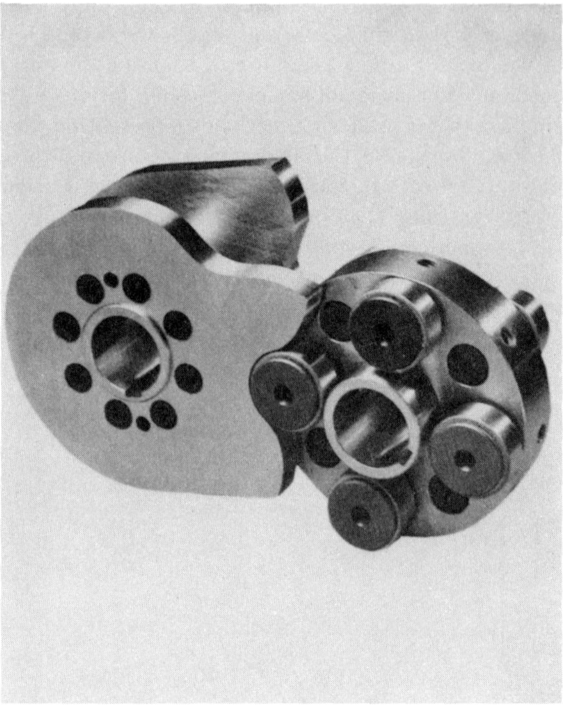

Fig. 1.3.16. Planar indexing cam mechanisms.
(Courtesy of Sankyo America, Inc.)

1.4 Design Considerations

The angle of rotation of the cam disk will be represented henceforth by ψ, while the resulting output function, whether denoting translational or rotational motion, is represented in this section by $f(\psi)$. A condition to be satisfied for any output function $f(\psi)$ is that it must be a continuous and periodic function of $\psi \in [0, 2\pi]$; that is,

$$f(\psi) = f(\psi + 2\pi) \tag{1.4.1}$$

If the foregoing condition is not observed, one could obtain cam curves, such as the ones shown in Fig. 1.4.1, which are unacceptable designs since, in general, such curves are unsuitable for applications requiring a constant angular velocity of the cam disk.

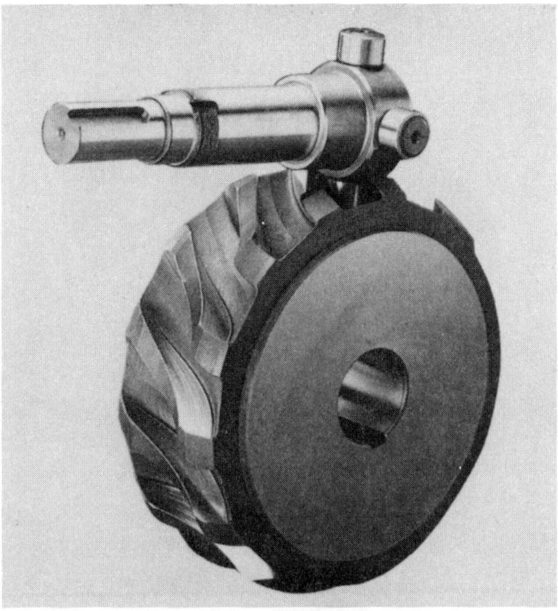

Fig. 1.3.17. Spatial indexing cam mechanism.
(Courtesy of Sankyo America, Inc.)

The condition introduced to avoid profiles similar to those of Fig. 1.4.1 is that ρ be a periodic function of θ, of period 2π. Moreover, the radius of curvature of the cam profile must be kept above a certain lower bound; otherwise, curves such as that shown in Fig. 1.4.1*a* arise, which produce the undesirable effect known as *undercutting*—see Chapter 5 for further discussion. Conditions similar to eq. (1.4.1) can be imposed on the velocity and acceleration. For example, from the periodicity condition, one has:

$$f(2\pi) - f(0) = 0 \tag{1.4.2}$$

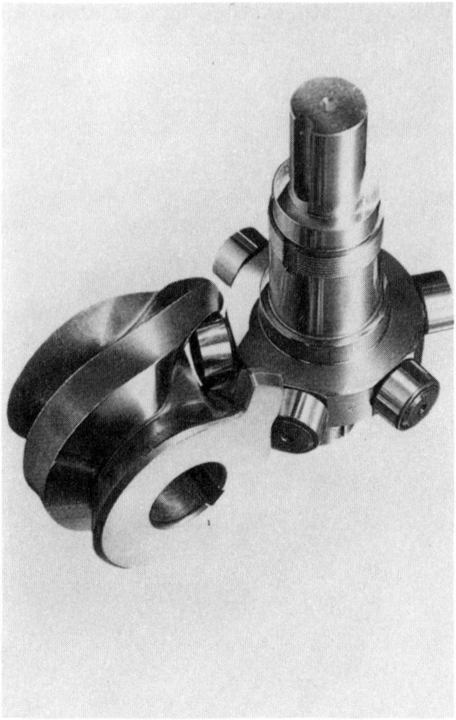

Fig. 1.3.18. Indexing cam mechanism with six indexing steps.
(Courtesy of Sankyo America, Inc.)

and, from the continuity condition,

$$\int_0^{2\pi} f'(\psi)\, d\psi = 0, \quad f'(\psi) \equiv \frac{df}{d\psi} \tag{1.4.3}$$

i.e., the area under the slope curve $f'(\psi)$, in the $f'(\psi)$ vs. ψ plot, must vanish. Furthermore, assuming a uniform cam velocity $\dot{\psi}(t)$, the acceleration of the follower can be written as

$$\ddot{f}(t) = f''(\psi)\, \dot{\psi}^2 \tag{1.4.4}$$

and hence, for the follower velocity to be periodic, one must have

$$\int_0^{2\pi} f''(\psi)\, d\psi = 0 \tag{1.4.5}$$

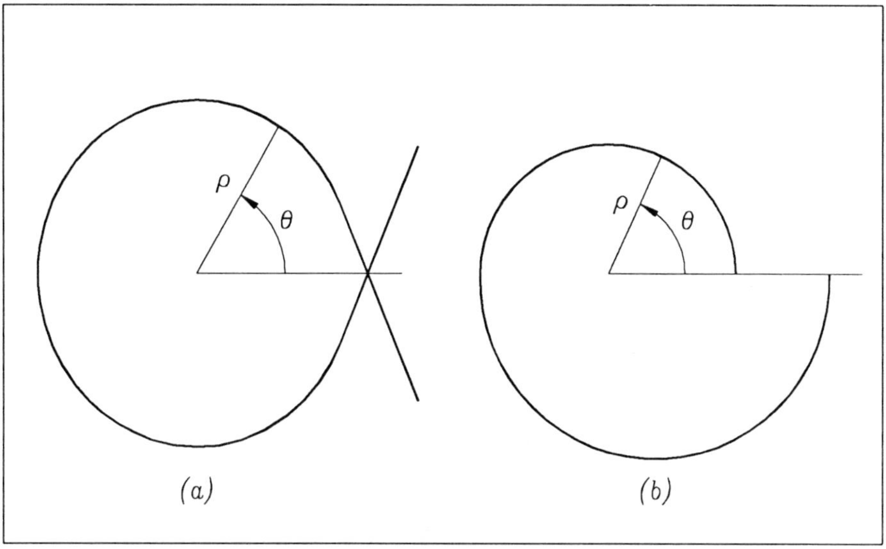

Fig. 1.4.1. Unacceptable curves for cam profiles.

i.e., the area under the *curvature* curve $f''(\psi)$ in the $f''(\psi)$ vs. ψ plot must vanish.

Another possibility of unsatisfactory design occurs when the cam curvature at some points attains inadmissibly large values. One way of eliminating this defect is by increasing the *base circle*. This is defined as the smallest circle centered at the cam axis of rotation and tangent to the cam profile. Nevertheless, the base circle cannot be increased without restriction, for the cam size grows correspondingly. In general, an optimum design is obtained when the cam is of minimum weight—i.e., of minimum area—and all functional constraints are satisfied. Three items affecting the cam-mechanism performance, namely, *the pressure angle*, *the eccentricity of the contact point*, and the *undercutting* phenomenon are discussed next.

1.4.1 Pressure Angle

The pressure angle is a function of merit in cam design. It appears in cams with both translating and oscillating knife-edge and roller followers, but we will limit our discussion to knife-edge followers. In fact, the pressure angle is defined for roller-followers based on the equivalent knife-edge follower, as discussed in detail in Chapter 5. The pressure angle in knife-edge followers, designated herafter as α, is defined as the angle between the normal to the cam profile at the contact point and the velocity of the contact point of the follower—see Figs. 1.4.2*a* & *b*.

In general, as the point of contact changes, due to the rotation of the cam, so does the pressure angle. Therefore, the latter is a function of the cam angular displacement ψ, i.e., $\alpha = \alpha(\psi)$. The pressure angle directly influences the force-transmission properties of the mechanism, as discussed in Chapter 5.

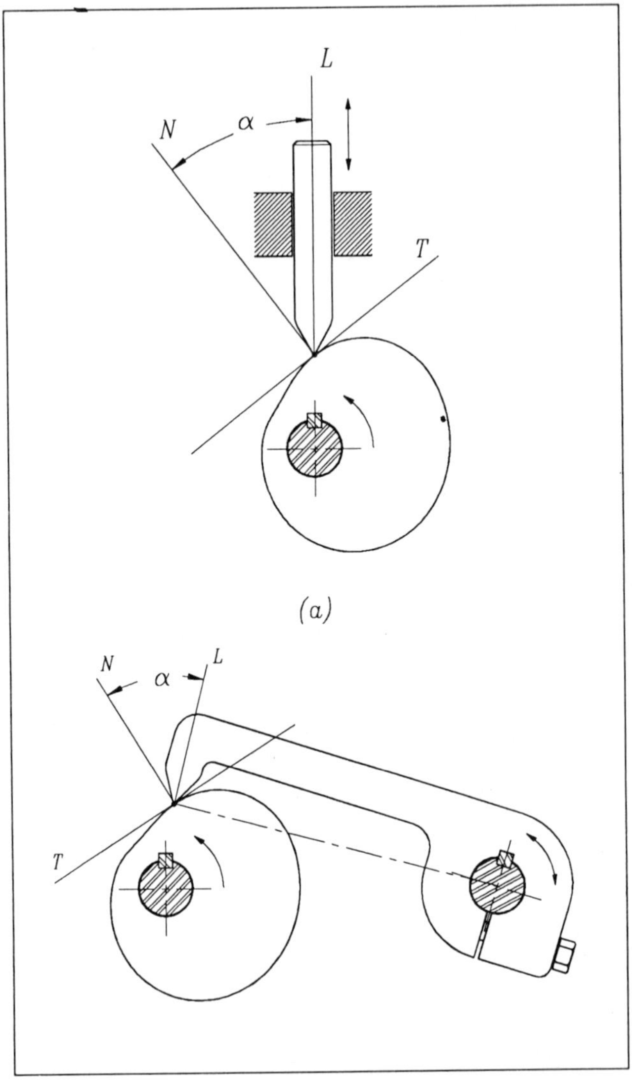

Fig. 1.4.2. Pressure angle of *a*) translating; and *b*) oscillating knife-edge follower.

The pressure angle determines how good the force transmission of the mechanism is. In fact, force transmission is better for small pressure angles (That is, regardless of the sign of α, the smaller the absolute value, the better the force transmission.) One way of keeping this value small is to increase the base circle. Nevertheless, as will be seen in Chapter 5, one can obtain the minimum base circle by imposing a constraint on the pressure angle. For a flat-face follower, the pressure angle becomes irrelevant, but other design concerns arise, as discussed below.

1.4.2 Design Considerations in Flat-Face Followers

In flat-face followers of either the translating or the roller types, the pressure angle is a constant and hence, irrelevant, because the follower can be designed so that this angle is maintained at a value of zero. Design considerations here are the *eccentricity of the contact point* and the appearance of *cusps* in the cam profile. The former, indicated as ϵ in Figs. 1.4.3*a* and *b*, denotes the location of the contact point on the follower from a reference point R of the same. Clearly, large values of ϵ imply a large follower, which in turn leads to larger inertia forces and larger space requirements of the overall mechanism. Therefore, in the design optimization of this type of follower, we aim at keeping the absolute value of ϵ as small as possible. This can be attained, again, as in the case of translating and oscillating knife-edge and roller followers, by diminishing the radius of the base circle. Nevertheless, as the radius of the base circle diminishes, *cusps* are likely to appear in the cam profile. These are undesirable because they produce unusually large contact stresses (Burr, 1981). Thus, while we will aim at keeping $|\epsilon|$ as small as possible, we will monitor the occurrence of cusps and design accordingly.

1.4.3 Undercutting

This item is a matter of concern in roller-follower cam mechanisms. Whether the cam profile be concave or convex, undercutting may be present. It appears when the follower radius, a, is greater than or equal to the minimum absolute value of the radius of curvature of the pitch curve, r_m—see Figs. 3.2.3 and 5.2.10. In the first case, i.e., if $a > r_m$, the desired follower motion cannot be produced. In the second case, i.e., $a = r_m$, cusps are produced on the cam profile, and inadmissibly large contact stresses develop. This situation ends in wear and the eventual shortening of the life span of the mechanism. Undercutting is taken into account for cam mechanism optimization by properly dimensioning the radius of the roller, as discussed in Chapter 5.

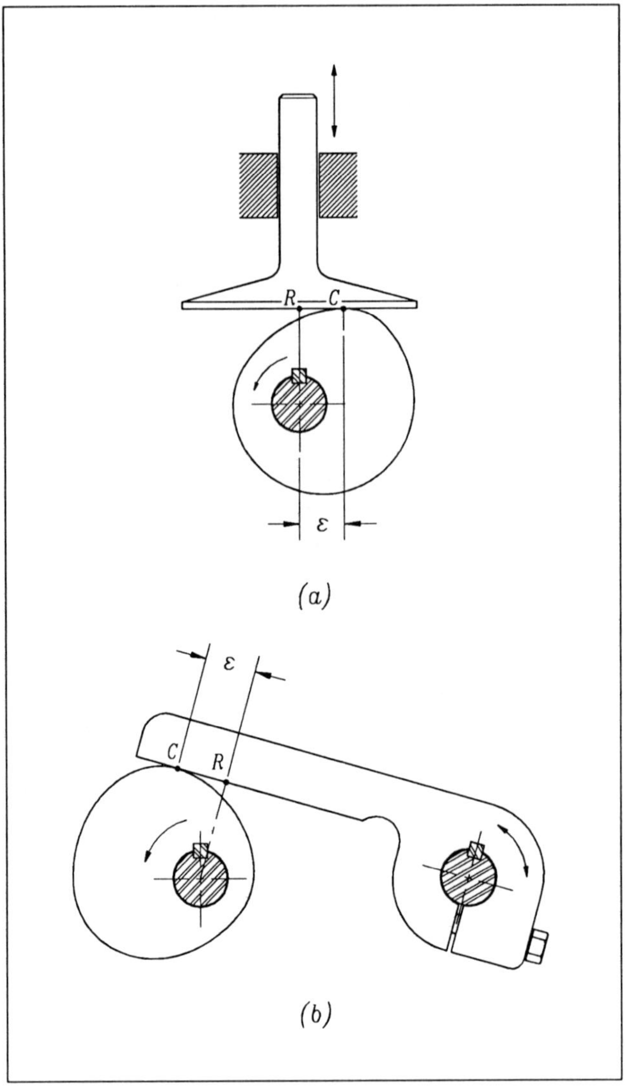

Fig. 1.4.3. Eccentricity of the contact point in:
a) the translating flat-face follower cam mechanism; and
b) the oscillating flat-face follower cam mechanism.

1.5 Scope of the Book

The displacement program is discussed in Chapter 2 where the problem definition is first presented. Then, conventional methods are included for completeness and, in the last section, spline methods are treated in detail. A review of cubic periodic splines is also included. In Chapter 3 the cam profile geometry, that is, the local and global geometric properties of cam profiles, are discussed. In Chapters 4 and 5, the optimum synthesis of planar cam mechanisms with flat-face and roller followers, respectively, is discussed, and Chapter 6 includes an introduction to spatial cam mechanisms through simple optimization techniques. These chapters form the core of the book. The kinematic analysis of cam mechanisms is outlined in Chapter 7 while, in Chapter 8, an outline of periodic parametric cubic splines, of interest for the analysis of the geometric properties of the cam profile, is included. Then, the interpolation of a given set of points in a plane with periodic parametric cubic splines producing a smooth closed contour, is presented in detail. Additionally, the computation of the global geometric properties of cam profiles is presented in the same chapter. In Chapter 9, an introduction is presented to the rendering of cams using a computer, whether on a monitor display, a hard copy, or the actual cam disk by means of a suitable machine tool. In Appendix A, the coefficients for the formulas derived in Chapter 8 are included. A set of review problems is included in Appendix B.

Chapter 2 THE DISPLACEMENT PROGRAM

2.1 Introduction

The synthesis of cam mechanisms to produce a given displacement program of the follower is divided into two stages. The first stage is meant to yield an easy-to-handle, smooth-enough function of the angle of rotation of the cam disk, representing the displacement of the follower. The second stage deals with the synthesis of the cam profile itself. The synthesis of the displacement program has been traditionally based on a limited set of functions giving rise to parabolic, harmonic, cycloidal, trapezoidal, and polynomial motions. Based on the last type of motions, an interesting method was proposed for the displacement-program synthesis using linear programming (Thompoulos and Knowles, 1975); recently, a totally new approach has been introduced using spline functions (Hitoshi, 1976; Nó Sánchez and García de Jalón, 1980; Angeles, 1983a). In this chapter, the different approaches for generating the displacement program are considered. The problem definition is stated in Section 2.2, and a review of conventional methods is presented in Section 2.3. A novel approach using spline methods is fully detailed in Section 2.4, and, in the last section of this chapter, the synthesis of the displacement program is outlined.

2.2 Problem Definition

The displacement program considered henceforth will be the most common one, namely, *dwell-rise-dwell-return*, $DRDR$, shown in Fig. 2.2.1. In that figure, we distinguish two dwell *phases*, the lower dwell and the upper dwell, which are denoted henceforth by D_1 and D_2, respectively. In addition, we distinguish a *rise phase* R_1 and a *return phase* R_2. Other follower motion programs, such as *rise-return* (RR) and *dwell-rise-return* (DRR), can be considered as special cases of the $DRDR$ program. Furthermore, more complex motions, such as those appearing in indexing cam mechanisms, can be synthesized using a combination of the basic elements constituting the $DRDR$ motion. In order to realize the rise phase of a cam-follower motion, it is necessary to synthesize a function $\sigma = \sigma(\psi)$ that connects the two dwell phases, D_1 and D_2, represented by lines parallel to the ψ axis in Fig. 2.2.1. In that figure, σ corresponds to the follower lift, measured from its lowest position, whereas ψ corresponds to the angle of rotation of the cam. Each of the foregoing phases of the *DRDR* motion program takes place in a subinterval $[\psi_1, \psi_2]$ of the cam motion interval $\psi \in [0, 2\pi]$. We adopt henceforth the convention indicated below to designate the subinterval in which each of these phases takes place:

- lower dwell D_1 in $[0, \psi_a]$;
- rise phase R_1 in $[\psi_a, \psi_b]$;
- upper dwell D_2 in $[\psi_b, \psi_c]$;
- return phase R_2 in $[\psi_c, 2\pi]$.

The curve R_1 connecting D_1 with D_2 should have both zero slope and zero curvature at $\psi = \psi_a$ and $\psi = \psi_b$. This is equivalent to saying that the follower velocity and acceleration are continuous at these points. Hence, in order to guarantee the continuity of both velocity and acceleration at $\psi = \psi_a$ and $\psi = \psi_b$, $\sigma'(\psi)$ and $\sigma''(\psi)$ should vanish at these points. It is desired, furthermore, that the acceleration change smoothly between ψ_a and ψ_b and, for this reason, a set of values of $\sigma''(\psi_i)$, denoted by σ_i'' are imposed on a set $\{\psi_i\}_1^n$, where

$$\psi_a = \psi_1 < \psi_2 < \psi_3 < \ldots < \psi_{n-1} < \psi_n = \psi_b$$

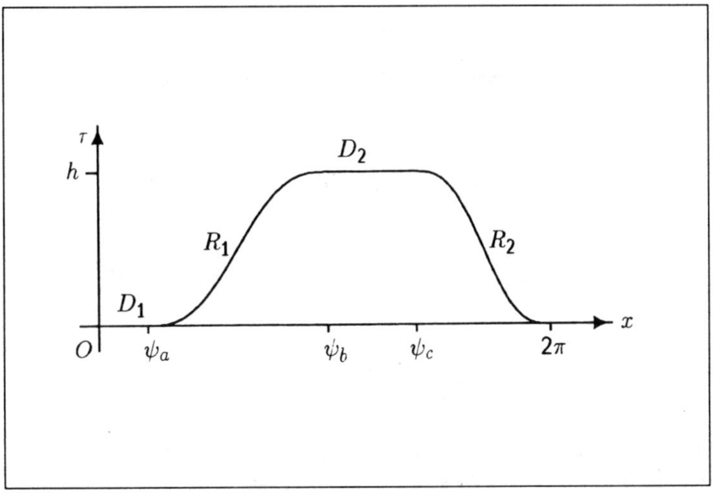

Fig. 2.2.1. Displacement program

A standard procedure widely applied in specifying the set $\{\sigma_i''\}_1^n$ has been to prescribe harmonic functions as follower displacement programs. These, however, are not suitable, for they present the drawback that at no point do their first and second derivatives vanish simultaneously. A solution to this problem has been to introduce either cycloidal or polynomial functions, as discussed in Section 2.3.

The discussion below will be made easier if we introduce a *normal motion program*. That is, for a motion taking place between the values ψ_a and ψ_b, producing a rise h of the follower, we define

$$\Delta\psi \equiv \psi_b - \psi_a > 0; \qquad x \equiv \frac{\psi - \psi_a}{\Delta\psi}, \quad \text{for} \quad \psi_a \leq \psi \leq \psi_b \tag{2.2.1a}$$

$$\sigma(\psi) = h\,\tau(x), \qquad 0 \leq x \leq 1 \tag{2.2.1b}$$

For a descent, the same applies, except that $\sigma(\psi)$ is now defined as

$$\sigma(\psi) = h\,\tau(1-x), \qquad 0 \leq x \leq 1 \tag{2.2.1c}$$

Therefore, $\tau(x)$ and x, as illustrated in Fig. 2.2.2, range between 0 and 1.

Henceforth, $\tau(x)$ is termed the *normal motion*. Note that the derivatives of $\tau(x)$ and $\sigma(\psi)$ are related by

$$\begin{aligned}
\sigma'(\psi) &= h\,x'(\psi)\,\tau'(x) = \frac{h}{\Delta\psi}\,\tau'(x)\\
\sigma''(\psi) &= \frac{h}{\Delta\psi}\,x'(\psi)\,\tau''(x) = \frac{h}{(\Delta\psi)^2}\,\tau''(x)\\
\sigma'''(\psi) &= \frac{h}{(\Delta\psi)^2}\,x'(\psi)\,\tau'''(x) = \frac{h}{(\Delta\psi)^3}\,\tau'''(x)\\
&\;\;\vdots\\
\sigma^{(n)}(\psi) &= \frac{h}{(\Delta\psi)^n}\,\tau^{(n)}(x)
\end{aligned} \tag{2.2.2}$$

where, obviously, the *chain rule* has been applied repeatedly.

In general, in order to synthesize motion programs, we will resort to some *tricks*, as described below. Let us assume that we have a curve represented by $f(x)$ as shown in Fig. 2.2.3.

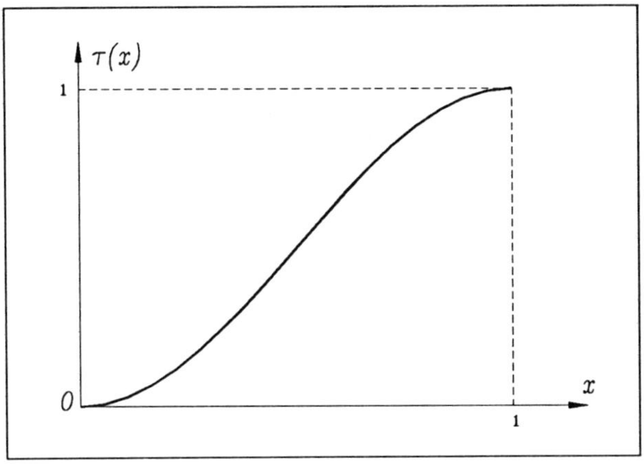

Fig. 2.2.2. Plot of the normal function $\tau(x)$

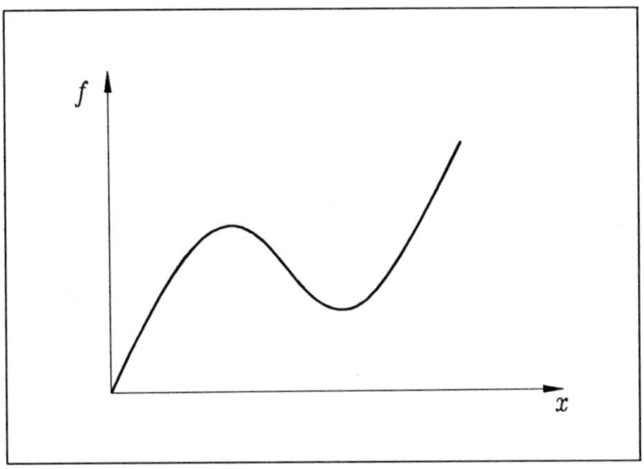

Fig. 2.2.3. Plot of the curve $f(x)$

One can mirror-image the function $f(x)$ about:

- the x-axis to obtain $-f(x)$ as shown in Fig. 2.2.4*a*; or
- the f-axis to obtain $f(-x)$ as shown in Fig. 2.2.4*b*.

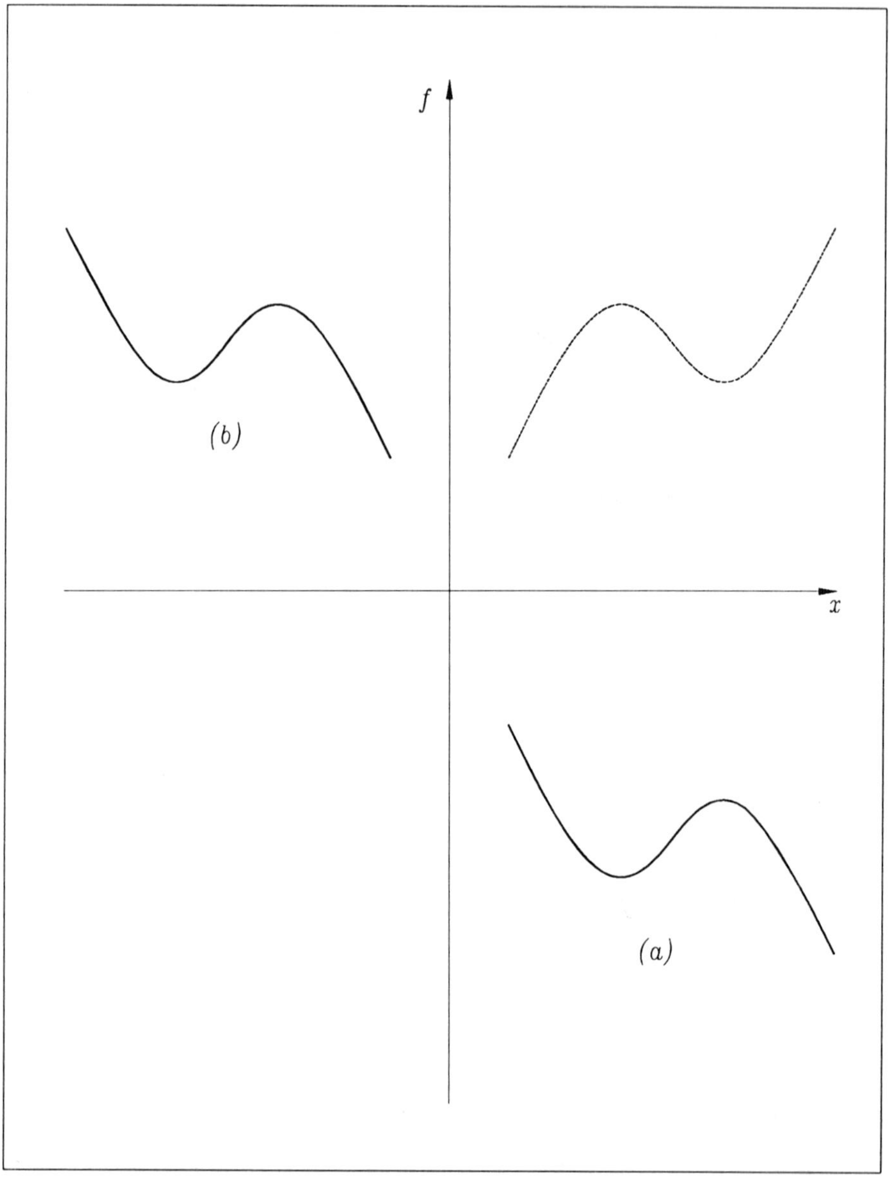

Fig. 2.2.4 Plot of the mirror image of $f(x)$ about: a) the x-axis; b) the f-axis

Moreover, one can also shift the curve $f(x)$ to a new origin:

- $O'(a,0)$, to obtain $f(x-a)$, as shown in Fig. 2.2.5*a*;
- $O'(0,b)$, to obtain $f(x)-b$, as shown in Fig. 2.2.5*b*; or
- $O'(a,b)$, to obtain $f(x-a)-b$, as shown in Fig. 2.2.6.

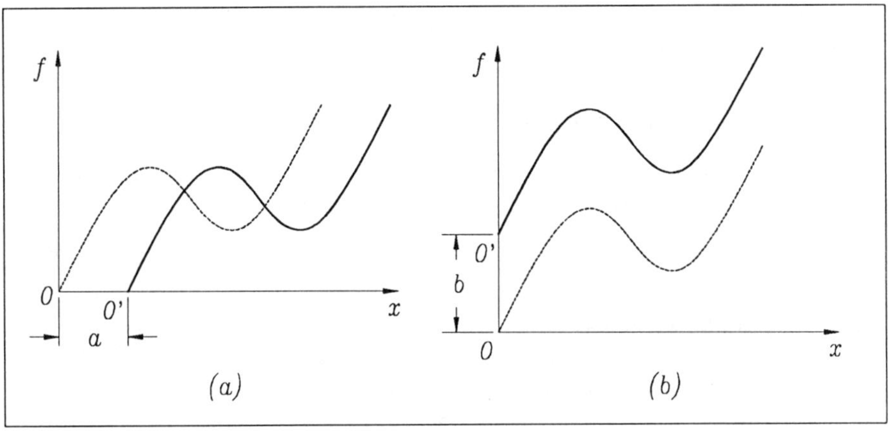

Fig. 2.2.5. Plot of the curve $f(x)$ shifted to origin: *a*) $O'(a,0)$; and *b*) $O'(0,b)$

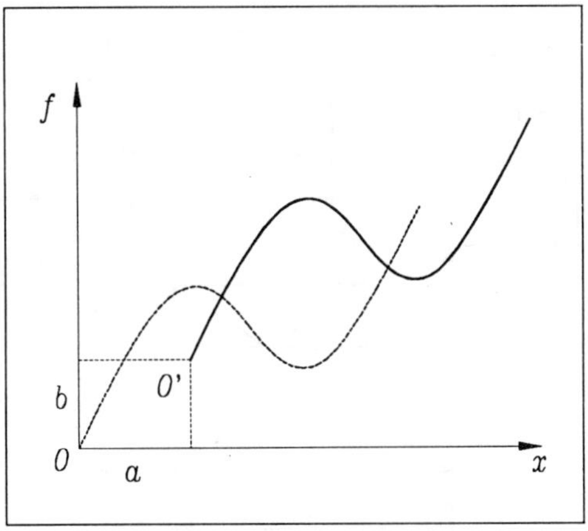

Fig. 2.2.6. Plot of the curve $f(x)$ shifted to the origin $O'(a,b)$.

2.3 Conventional Methods

In this section, a brief outline of the conventional methods of displacement-program synthesis is presented. The kinematic equations for each program are stated and, in some cases, formally derived. However, given the extensive literature already written on the subject (Rothbart, 1956; Chen, 1982), graphical methods for constructing the actual curves are omitted. In fact, graphical methods of constructing harmonic or cycloidal functions, for example, have become anachronisms in light of the hardware and software readily available for plotting. In each case, the limitations of such methods to the synthesis of the cam profile are highlighted.

2.3.1 Constant-Velocity Program

The simplest kind of follower motion corresponds to that of constant velocity. In this case the normal motion, τ, is expressed simply as

$$\tau(x) = x \tag{2.3.1}$$

Thus, $\sigma(\psi)$ can be written as

$$\sigma(\psi) = h\,\tau(x) = h\,\frac{\psi - \psi_a}{\Delta\psi}, \qquad \psi_a \le \psi \le \psi_b \tag{2.3.2a}$$

Therefore,

$$\sigma'(\psi) = \frac{d\sigma}{d\psi} = \frac{h}{\Delta\psi}, \qquad \psi_a \le \psi \le \psi_b \tag{2.3.2b}$$

and

$$\sigma''(\psi) = \frac{d^2\sigma}{d\psi} = \begin{cases} 0, & \text{for } \psi_a < \psi < \psi_b; \\ \infty, & \text{for } \psi = 0,\ \psi_b \end{cases} \tag{2.3.2c}$$

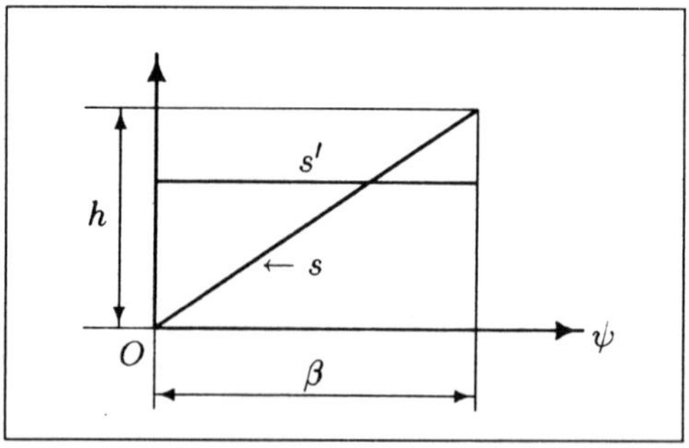

Fig. 2.3.1. Constant-velocity program

The big disadvantage of this program lies in the velocity discontinuities at the blending points $\psi = \psi_a$ and $\psi = \psi_b$, i.e., points at which the function $\sigma(\psi)$, defined in eq. (2.3.2a), joins the constant-displacement lines $\sigma = 0$ and $\sigma = h$. Clearly, at the blending points, jump discontinuities of $\sigma'(\psi)$ and hence, infinite discontinuities of $\sigma''(\psi)$ arise. Therefore, the usage of this program requires the introduction of smoothing polynomials at the blending points—see Problems B.1 and B.2 in Appendix B.

2.3.2 Constant-Acceleration Program

In this program, the normal motion $\tau(x)$, can be expressed as a quadratic function of x, i.e.,

$$\tau(x) = x^2 \qquad 0 \le x \le 1 \tag{2.3.3a}$$

and

$$\tau'(x) = 2x, \quad \tau''(x) = 2 \qquad 0 \le x \le 1 \tag{2.3.3b}$$

This motion program is shown in Fig. 2.3.2.

2.3.3 Piecewise Constant-Acceleration Program

This motion program is expressed as:

$$\tau(x) = \begin{cases} 2x^2, & \text{for } 0 \le x \le \frac{1}{2}; \\ 1 - 2(x-1)^2, & \text{for } \frac{1}{2} \le x \le 1 \end{cases} \tag{2.3.4}$$

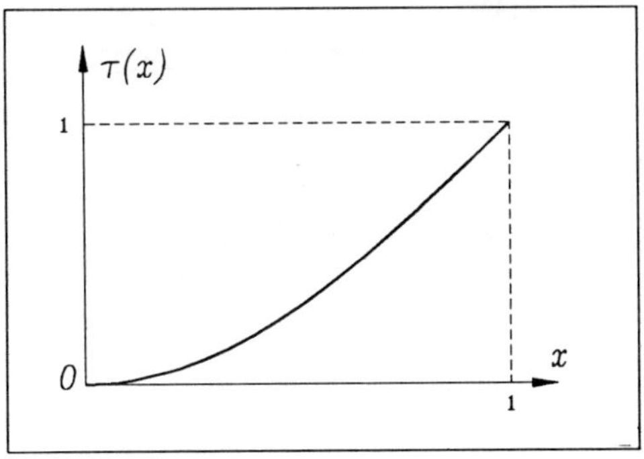

Fig. 2.3.2. Constant acceleration program

We synthesize the foregoing motion as outlined below:

- by mirror-imaging $\tau(x)$—see Fig. 2.3.2—about the x and τ axes, the curve shown in Fig. 2.3.3a is obtained,
- now the mirror-imaged curve is translated to the new origin $O'(1,1)$, as shown in Fig. 2.3.3b.
- the desired motion program is displayed in Fig. 2.3.4.

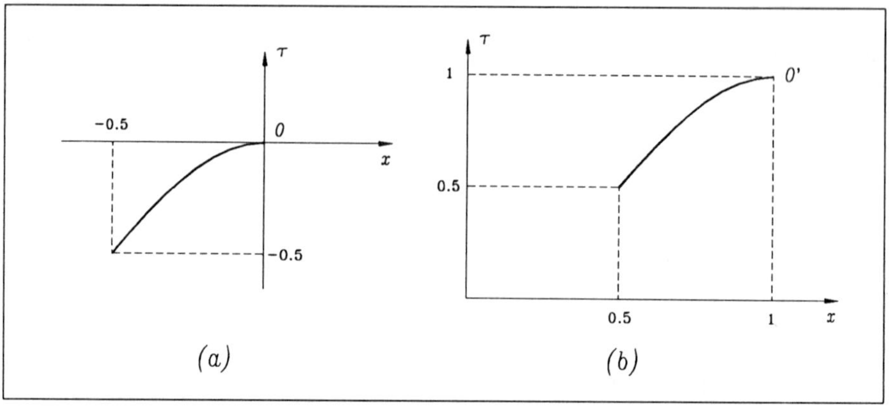

Fig. 2.3.3. Curve $\tau(x)$: a) mirror-imaged about the x and τ axes; b) mirror-imaged $\tau(x)$ translated to a new origin $O'(1,1)$

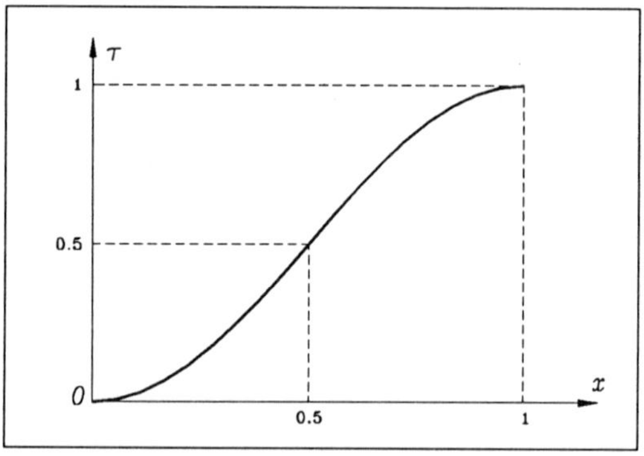

Fig. 2.3.4. Piecewise constant-acceleration program

For both types of motion programs, however, the third derivative of τ with respect to x, i.e., $\tau'''(x)$, and hence, the follower jerk, become infinitely large at points where the acceleration undergoes a jump discontinuity, namely, at $x = 0$, $x = 1/2$, and $x = 1$. Because of this effect, the usage of this program is very limited as such. For practical implementation, the introduction of blending functions at the jump discontinuities is required.

2.3.4 Simple Harmonic Motion

The normal simple harmonic motion and its derivatives are expressed as:

$$\tau(x) = \frac{1}{2}(1 - \cos \pi x), \qquad 0 \le x \le 1 \tag{2.3.5a}$$

$$\tau'(x) = \frac{\pi}{2} \sin \pi x, \qquad 0 \le x \le 1 \tag{2.3.5b}$$

$$\tau''(x) = \frac{\pi^2}{2} \cos \pi x, \qquad 0 \le x \le 1 \tag{2.3.5c}$$

Plots of the foregoing functions are shown in Fig. 2.3.5. It can be seen from those plots that, at the blending points $x = 0$ and $x = 1$, there are acceleration jumps that will produce infinite jerk values at those points. The use of this program is therefore limited to cases in which acceleration jumps can be tolerated.

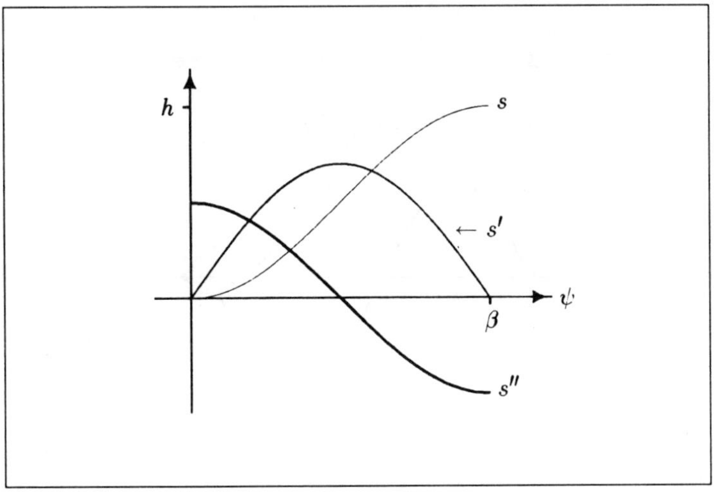

Fig. 2.3.5. Simple harmonic motion

2.3.5 Cycloidal Motion

In this case, the normal motion and its derivatives are expressed as:

$$\tau(x) = x - \frac{1}{2\pi}\sin 2\pi x, \qquad 0 \leq x \leq 1 \tag{2.3.6a}$$

$$\tau'(x) = 1 - \cos 2\pi x, \qquad 0 \leq x \leq 1 \tag{2.3.6b}$$

$$\tau''(x) = 2\pi \sin 2\pi x, \qquad 0 \leq x \leq 1 \tag{2.3.6c}$$

Functions (2.3.6*a*–*c*) are plotted in Fig. 2.3.6. Since there are no discontinuities in the acceleration at $x = 0$ and $x = 1$, this motion program is widely used in high-speed applications. Nevertheless, conditions for acceleration can only be satisfied at the blending points. Should acceleration conditions be required at points other than the blending points, the designer would have to resort to another type of function, as explained in Subsection 2.3.6.

Note that the cycloidal motion produces a maximum *normal velocity* $\tau'(x)$ of 2 at $x = 0.5$, whereas its maximum and minimum *normal acceleration* $\tau''(x)$ occur at $x = 0.25$ and $x = 0.75$, respectively, both with the same absolute value, namely, 2π.

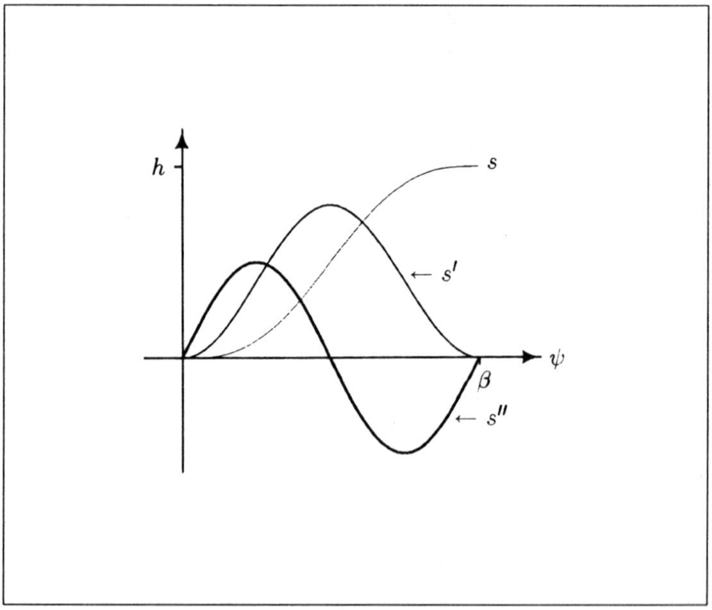

Fig. 2.3.6. Cycloidal motion

So far, we have studied the so-called "basic curves" for the follower-displacement program. Those curves were obtained either from simple polynomials or from trigonometric functions. Combination of those curves give rise to "modified curves", the objective of the latter being to produce "smoothness" at the blending points. In Subsection 2.3.6, a short review of higher-degree polynomials and orthogonal polynomials is presented.

2.3.6 Follower-Displacement Programs Generated by Polynomials

In principle, if there are $n+1$ conditions on displacement, velocity, and acceleration to be met, one can join the lower and upper dwells with a polynomial of degree n. That is, the normal motion $\tau(x)$, can be expressed as:

$$\tau(x) = a_0 + a_1 x + a_2 x^2 + \ldots + a_n x^n \tag{2.3.7}$$

where the constant coefficients a_i, for $i = 0, \ldots, n$, are determined from the conditions to be satisfied. Next, an outline of the method followed to determine the polynomial coefficients is introduced. One is interested in obtaining a polynomial whose kth derivative attains a prescribed value S_{kj} at $x = x_j$, where $j, k = 0, 1, \ldots, n$. This condition can be stated as

$$k!\, a_k + (k+1)!\, a_{k+1} x_j + \cdots + \frac{n!}{(n-k)!} a_n\, x_j^{n-k} = S_{kj} \tag{2.3.8}$$

From eq. (2.3.8) it is apparent that the unknowns $\{a_i\}_0^n$ appear linearly in those equations. Thus, in general, a system of linear equations is derived, namely,

$$\mathbf{M}\,\mathbf{a} = \mathbf{b} \tag{2.3.9}$$

where $\mathbf{M}$ is an $(n+1) \times (n+1)$ known matrix, $\mathbf{b}$ is an $(n+1)$-dimensional known vector, and

$$\mathbf{a} = [\, a_0,\ a_1,\ \ldots,\ a_n\,]^T$$

is the $(n+1)$-dimensional vector of unknowns. However, as the degree of the polynomial increases, so do the dimension and the sensitivity of the linear system of equations to be solved, this system thus becoming ill-conditioned. Ill-conditioning means that a "large" value of the *condition number* (Golub and Van Loan, 1983) of matrix $\mathbf{M}$ arises. The presence of a condition number of, say, 10^6, may be inadmissible because the said number measures the amplification of the roundoff (relative) error of the results with respect to the roundoff (relative) error of the data.

To illustrate the effect of ill-conditioning, let us assume that we want to find the coefficients $\{a_i\}_0^n$ of an nth-order polynomial attaining the values $\{y_i\}_0^n$ at the abscissae $\{x_i\}_0^n$ where

$$x_i \equiv \frac{i}{n}, \qquad i = 0, \ldots, n \tag{2.3.10}$$

In this case, matrix $\mathbf{M}$ and vector $\mathbf{b}$ take on the forms

$$\mathbf{M} \equiv \begin{bmatrix} 1 & x_0 & x_0^2 & \dots & x_0^n \\ 1 & x_1 & x_1^2 & \dots & x_1^n \\ \vdots & \vdots & \vdots & \ddots & \vdots \\ 1 & x_n & x_n^2 & \dots & x_n^n \end{bmatrix}, \quad \mathbf{b} = \begin{bmatrix} y_o \\ y_1 \\ \vdots \\ y_n \end{bmatrix} \tag{2.3.11}$$

The condition number κ of matrix $\mathbf{M}$ as shown above, for various values of n, obtained with *MATLAB* (1990), are given in Table 2.3.1 below.

Table 2.3.1

n	2	3	4	5	10	15
κ	16	99	686	4924	9.64×10^7	1.77×10^{10}

As observed in Table 2.3.1, the condition number of the matrix under study becomes greater than one million for $n \geq 10$. This means that, upon solving eq. (2.3.9) for $\mathbf{a}$, the results will be corrupted with a relative roundoff error greater than one million times the relative roundoff error of the data. This roundoff error amplification may be inadmissible in some instances.

In order to avoid the ill-conditioning problem, some displacement programs are generated by resorting to *orthogonal polynomials*. These are represented as:

$$\tau(x) = k_0\, P_0(x) + k_1\, P_1(x) + k_2\, P_2(x) + \ldots + a_m\, P_m(x) \tag{2.3.12}$$

where k_i and $P_i(x)$, for $i = 0, \ldots, m$ are, respectively, constant coefficients and a family of orthogonal polynomials. The main feature of these polynomials is that they behave as orthogonal coordinates in the sense of yielding better numerical results than non-orthogonal polynomials.

An important example of orthogonal polynomials are the *Chebyshev* polynomials defined by

$$T_j(x) = \cos(j\,\varphi), \quad \varphi = \cos^{-1}(x), \quad -1 \leq x \leq 1$$

for $j = 0, 1, 2, \ldots, n$. For instance, for $n = 4$, the aforementioned polynomials are

$$T_0(x) = 1$$

$$T_1(x) = x$$

$$T_2(x) = 2\,x^2 - 1$$

$$T_3(x) = 4\,x^3 - 3\,x$$

$$T_4(x) = 8\,x^4 - 8\,x^2 + 1$$

In the above set, some of many useful properties of Chebyshev polynomials listed below can be easily verified, namely,

- they can be computed through a recursive formula, i.e.,

$$T_0(x) = 1, \quad T_{j+1}(x) = 2\,x\,T_j(x) - T_{j-1}(x), \quad j \geq 1$$

- the leading coefficient is 2^{j-1} for $j \geq 1$ and 1 for $j = 0$.
- they have the symmetry property, namely, $T_j(-x) = (-1)^j\,T_j(x)$
- $T_j(x)$ has j zeros in $[-1,\ 1]$ given by

$$x_k = \cos\left(\frac{\pi}{2}\,\frac{2\,k+1}{j}\right), \quad k = 0, 1, 2, \ldots, j-1$$

  These are the *Chebyshev abscissae*. Moreover, they have $j + 1$ extrema given as

$$x_k^* = \cos\frac{k\,\pi}{j}, \quad T_j(x_k^*) = (-1)^k, \quad k = 0, 1, 2, \ldots, j$$

- they have the orthogonality property, i.e., given two functions of x, f and g, and their inner product defined as

$$(f,\, g) = \sum_{k=0}^{m} f(x_k)\,g(x_k),$$

  where x_k are the zeros of $T_{m+1}(x)$, then, for $0 \leq i \leq m$, $0 \leq l \leq m$, one has

$$(T_i,\, T_l) = \begin{cases} 0 & \text{if } i \neq l \\ (m+1)/2 & \text{if } i = l \neq 0 \\ m+1 & \text{if } i = l = 0 \end{cases}$$

- Minimax property: Of all jth-degree polynomials with leading coefficient 1, $2^{1-j}T_j$ has the smallest maximum norm in $[-1,\ 1]$, this maximum norm being 2^{1-j}.

A comprehensive discussion of orthogonal polynomials in the realm of numerical analysis is presented in (Dahlquist and Björk, 1974). Moreover, for a detailed discussion of displacement programs based on this type of polynomials and many others, such as programs based on osculating polynomials, modified parabolic motion, modified trapezoidal and skewed trapezoidal, Fourier series curves, Polydyne curves, etc., and combinations thereof, the interested reader is referred to Chen (1982), Chakraborty and Dhande (1977), Tesar and Matthew (1976), and Rothbart (1956). An example of a motion program known as the 3-4-5 polynomial is presented below.

2.3.7 Example: A 3-4-5 Polynomial Motion

This motion is interesting because it can produce a smooth blending at its end points. That is, it verifies six end conditions, namely,

$$\tau(0) = \tau'(0) = \tau''(0) = 0, \qquad \tau(1) = 1, \quad \tau'(1) = \tau''(1) = 0 \tag{2.3.13}$$

Since we have six conditions to satisfy, we need a fifth degree polynomial, i.e.,

$$\tau(x) = k_0 + k_1 x + k_2 x^2 + k_3 x^3 + k_4 x^4 + k_5 x^5 \tag{2.3.14}$$

and its first two derivatives, namely,

$$\begin{aligned} \tau'(x) &= k_1 + 2k_2 x + 3k_3 x^2 + 4k_4 x^3 + 5k_5 x^4 \\ \tau''(x) &= 2\,k_2 + 6\,k_3 x + 12\,k_4 x^2 + 20\,k_5 x^3 \end{aligned} \tag{2.3.15}$$

Substitution of the conditions to be satisfied at $x = 0$ and $x = 1$ yields

$$\begin{aligned} \tau(0) &= 0 = k_0 \\ \tau'(0) &= 0 = k_1 \\ \tau''(0) &= 0 = k_2 \\ \tau(1) &= 1 = k_0 + k_1 + k_2 + k_3 + k_4 + k_5 \\ \tau'(1) &= 0 = k_1 + 2k_2 + 3k_3 + 4k_4 + 5k_5 \\ \tau''(1) &= 0 = 2k_2 + 6k_3 + 12k_4 + 20k_5 \end{aligned} \tag{2.3.16}$$

Clearly, $k_0 = k_1 = k_2 = 0$. Therefore, the system of linear equations to be solved is

$$\begin{bmatrix} 1 & 1 & 1 \\ 3 & 4 & 5 \\ 6 & 12 & 20 \end{bmatrix} \begin{bmatrix} k_3 \\ k_4 \\ k_5 \end{bmatrix} = \begin{bmatrix} 1 \\ 0 \\ 0 \end{bmatrix} \tag{2.3.17}$$

the solution being

$$k_3 = 10, \qquad k_4 = -15, \qquad k_5 = 6$$

and hence, $\tau(x)$ turns out to be of the form

$$\tau(x) = 10\,x^3 - 15\,x^4 + 6\,x^5 = x^3\,(10 - 15\,x + 6\,x^2) \tag{2.3.18}$$

from which it is apparent why this is called a *3-4-5 polynomial motion*. Its derivatives are displayed below for quick reference:

$$\begin{aligned} \tau'(x) &= 30\,(x^2 - 2x^3 + x^4) = 30\,x^2\,(x-1)^2 \\ \tau''(x) &= 60\,(x - 3x^2 + 2x^3) = 60\,x\,(x-1)\,(2x-1) \\ \tau'''(x) &= 60\,(1 - 6x + 6x^2) \\ \tau^{iv}(x) &= -360\,(1-2x) \\ \tau^{v}(x) &= 720 \\ \tau^{vi}(x) &= 0 \end{aligned} \tag{2.3.19}$$

Note that $\tau'(x)$ vanishes at the ends of the interval $[\,0,\ 1\,]$, which are then local minima of this function, a local maximum, 1.875, occurs at $x = 0.5$. Moreover, the second derivative attains a local maximum of 5.7736 at $x_1 = 0.2113$ and a local minimum of -5.7736 is reached at $x_2 = 0.7887$, and $x_1 + x_2 = 1$. Thus, the maximum velocity and acceleration of the 3-4-5 polynomial motion are lower than those of the cycloidal motion, a rather interesting result. In fact, the maximum velocity of the 3-4-5 polynomial is 93.75% of the corresponding value for the cycloidal motion, whereas the maximum acceleration of the 3-4-5 polynomial motion is 91.89% of the same value for the cycloidal motion. Let $\tau_P(x)$ and $\tau_C(x)$ denote the 3-4-5 polynomial and the cycloidal motions, respectively. The differences $\tau_C(x) - \tau_P(x)$ and their first two derivatives with respect to x are plotted in Figs. 2.3.7–2.3.9, respectively.

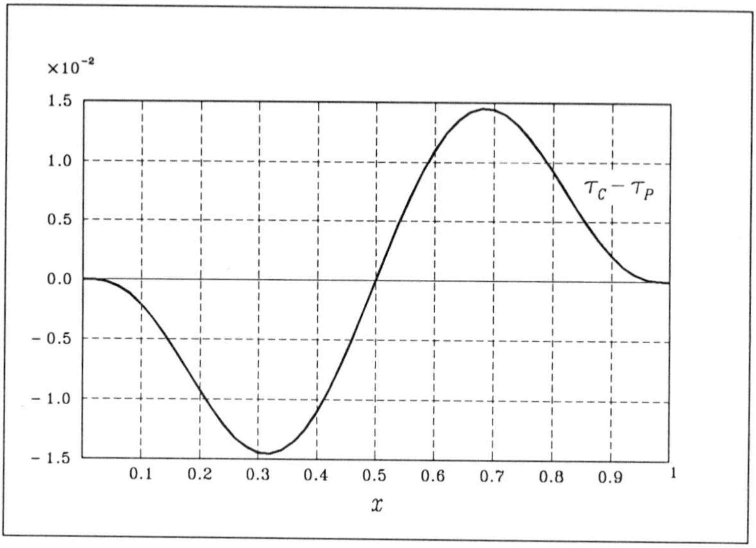

Fig. 2.3.7. Difference in displacement between the 3-4-5 polynomial and cycloidal motions.

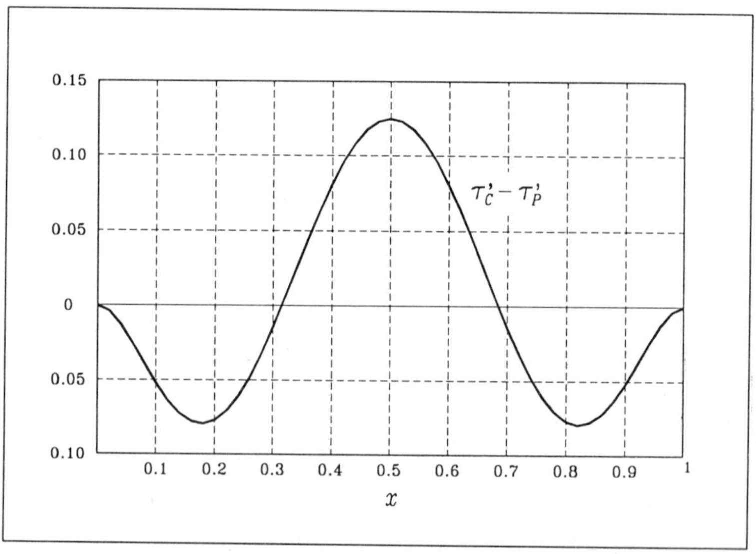

Fig. 2.3.8. Difference in velocity between the 3-4-5 polynomial and cycloidal motions.

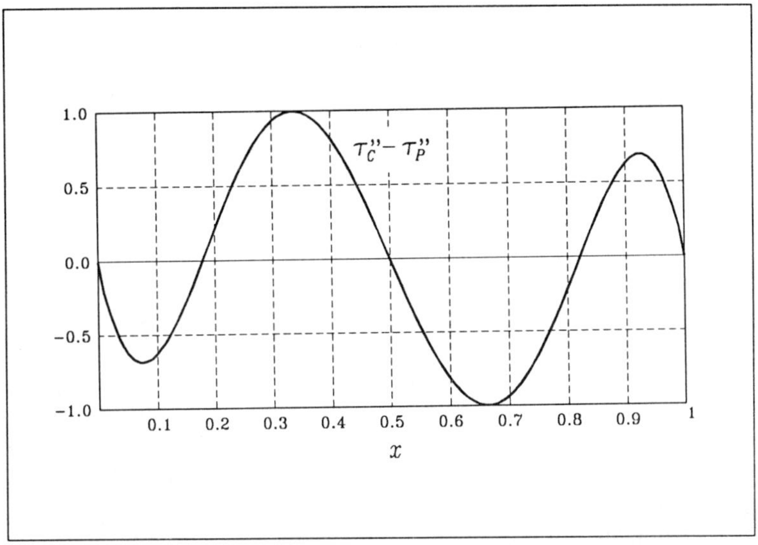

Fig. 2.3.9. Difference in acceleration between the 3-4-5 polynomial and cycloidal motions.

2.4 Spline Methods

As discussed in the foregoing sections, the generation of curves for motion programs, satisfying a set of ordinate and derivative conditions, has led to the use of several kinds of polynomials. Nevertheless, even if these are chosen to be smooth enough, in the sense of being maximally differentiable at the blending points, e.g., Lagrange interpolating polynomials, as the number of points increases, these polynomials can oscillate strongly. As a consequence, the trend in recent years has been to resort to spline curves composed of lower-degree polynomials for the generation of such motion programs (Hitoshi, 1976; Nó-Sánchez and García de Jalón, 1980; Angeles, 1983a). Spline curves allow the handling of many points or conditions which, in turn, ease their control. In Section 2.4.1, an introduction to cubic spline functions is outlined, whereas in Section 2.4.2, periodic cubic spline functions are discussed. Finally, the generation of the displacement program using the normal motion is presented in Subsection 2.4.3.

2.4.1 Cubic Splines

A cubic spline function $s(x)$ connecting n points P_k $(x_k,\ y_k)$, for $k = 1, 2, \ldots, n$, is a curve composed of $n-1$ third-degree polynomials joined at the points P_k so that

$s(x_k) = y_k$. Moreover, they are twice differentiable at those points. Hence, cubic splines are said to be C^2 curves, i.e., to have continuous derivatives up to the second order.

Cubic splines are optimal in the sense that they minimize a *functional*, i.e., an integral defined as

$$F = \int_0^T s''^2(x)\, dx \tag{2.4.1a}$$

subject to the constraints

$$s(x_k) = y_k, \qquad k = 1, \ldots, n \tag{2.4.1b}$$

where x_k and y_k are given. The foregoing optimality property has a simple kinematic interpretation: among all curves passing through a set of points $P_1, P_2, \ldots, P_n$ in the x-y plane, the cubic spline is the one containing the minimum *acceleration magnitude*. In fact, F, as given above, is the square of the Euclidean norm (Halmos, 1974) of $s''(x)$, i.e., F turns out to be a measure of the magnitude of the acceleration of a displacement program given by $s(x)$.

Let $P_k(x_k, y_k)$ and $P_{k+1}(x_{k+1}, y_{k+1})$ be two consecutive supporting points. The kth cubic polynomial $s_k(x)$ between those points is given by:

$$s_k(x) = A_k\,(x - x_k)^3 + B_k\,(x - x_k)^2 + C_k\,(x - x_k) + D_k, \quad x_k \le x \le x_{k+1} \tag{2.4.2a}$$

Thus, for the spline $s(x)$, $4\,(n-1)$ coefficients—A_k, B_k, C_k, D_k, for $k = 1, \ldots, n-1$—are to be determined. They are readily computed in terms of the function values and second derivatives, namely (Späth, 1978),

$$A_k = \frac{1}{6\,\Delta x_k}\,(s''_{k+1} - s''_k) \tag{2.4.2b}$$

$$B_k = \frac{1}{2}\, s''_k \tag{2.4.2c}$$

$$C_k = \frac{\Delta s_k}{\Delta x_k} - \frac{1}{6}\,\Delta x_k\,(s''_{k+1} + 2s''_k) \tag{2.4.2d}$$

$$D_k = s_k \tag{2.4.2e}$$

with

$$\Delta x_k = x_{k+1} - x_k, \qquad \Delta s_k = s_{k+1} - s_k \tag{2.4.2f}$$

s_k being the ordinate value at $x = x_k$, i.e., $s_k \equiv s(x_k)$, and $k = 1, \ldots, n-1$. Further definitions are $s'_k \equiv s'(x_k)$ and $s''_k \equiv s''(x_k)$.

Furthermore, the requirement of continuity in the first derivative is imposed, namely,

$$s'_{k-1}(x_k) = s'_k(x_k), \quad k = 2, \ldots, n-1, \qquad s' \equiv \frac{ds}{dx}$$

After substituting eqs. (2.4.2*b*–*f*) into eq. (2.4.2*a*), one obtains a linear system of $n-2$ simultaneous equations for the n unknowns $\{s''_k\}_1^n$.

Let $\mathbf{s}$ be the n-dimensional vector whose kth component is s_k, vector $\mathbf{s}''$ being defined likewise, i.e.,

$$\mathbf{s} = [\, s_1, \cdots, s_n \,]^T, \qquad \mathbf{s}'' = \left[\, s''_1, \cdots, s''_n \,\right]^T \tag{2.4.3a}$$

The relationship between $\mathbf{s}$ and $\mathbf{s}''$ (Späth, 1978) can be written in vector form as

$$\mathbf{A}\,\mathbf{s}'' = 6\,\mathbf{C}\,\mathbf{s} \tag{2.4.3b}$$

where $\mathbf{A}$ and $\mathbf{C}$ are $(n-2) \times n$ matrices defined as:

$$\mathbf{A} = \begin{bmatrix} \alpha_1 & 2\alpha_{1,2} & \alpha_2 & 0 & \cdots & 0 & 0 \\ 0 & \alpha_2 & 2\alpha_{2,3} & \alpha_3 & \cdots & 0 & 0 \\ \vdots & \vdots & \ddots & \ddots & \ddots & \vdots & \vdots \\ 0 & 0 & \cdots & \alpha_{n'''} & 2\alpha_{n''',n''} & \alpha_{n''} & 0 \\ 0 & 0 & 0 & \cdots & \alpha_{n''} & 2\alpha_{n'',n'} & \alpha_{n'} \end{bmatrix} \tag{2.4.3c}$$

and

$$\mathbf{C} = \begin{bmatrix} \beta_1 & -\beta_{1,2} & \beta_2 & 0 & \cdots & 0 & 0 \\ 0 & \beta_2 & -\beta_{2,3} & \beta_3 & \cdots & 0 & 0 \\ \vdots & \vdots & \ddots & \ddots & \ddots & \vdots & \vdots \\ 0 & 0 & \cdots & \beta_{n'''} & -\beta_{n''',n''} & \beta_{n''} & 0 \\ 0 & 0 & 0 & \cdots & \beta_{n''} & -\beta_{n'',n'} & \beta_{n'} \end{bmatrix} \tag{2.4.3d}$$

while

$$\alpha_k \equiv \Delta x_k, \quad \alpha_{i,j} \equiv \alpha_i + \alpha_j, \quad \beta_k \equiv 1/\alpha_k, \quad \beta_{i,j} \equiv \beta_i + \beta_j; \quad i, j, k = 1, \ldots, n' \tag{2.4.3e}$$

and

$$n' \equiv n-1, \qquad n'' \equiv n-2, \qquad n''' \equiv n-3 \tag{2.4.3f}$$

Thus, two additional equations are needed to render eq. (2.4.3b) a determined system. The additional equations are derived, in turn, depending upon the class of functions one is dealing with, which thus gives rise to various types of splines. For example, if we define s_1'' and s_n'' as zero, then we obtain *natural cubic splines*. On the other hand, if one is interested in periodic functions, which is the case when synthesizing follower-displacements motions, then the conditions $s_1 = s_n$, $s_1'' = s_n''$ are imposed, thereby producing *periodic cubic splines*.

In the subsection below, it will become apparent that the displacement program $\sigma(\psi)$ undergone by cam followers can be synthesized with the sole aid of periodic cubic splines and constant functions to account for dwells.

2.4.2 Periodic Cubic Splines

In this subsection, we illustrate the use of periodic cubic splines for the synthesis of the follower motion. We first synthesize a *normal motion* defined over abscissa and ordinate intervals $[0,\ 1]$. Based on this motion, the displacement program will be synthesized so that the rise and return phases will be produced by simple scaling and translating of the normal motion.

In the rise phase, a set of end conditions to satisfy are shown below:

$$\tau(0)=0, \quad \tau'(0)=0, \quad \tau''(0)=0 \tag{2.4.4}$$

and

$$\tau(1)=1, \quad \tau'(1)=0, \quad \tau''(1)=0 \tag{2.4.5}$$

The end conditions for function $\tau(x)$ present no difficulty; those for its first and second derivative, however, are not inherent to all cubic splines. Nevertheless, the foregoing conditions can all be met by defining $\tau(x)$ as periodic, of period 2, and by imposing that it be symmetric with respect to $\tau = 1$, i.e.,

$$\tau(x) = \tau(x+2), \qquad \tau(x) = \tau(2-x) \qquad 0 \le x \le 2 \tag{2.4.6}$$

Moreover, it is desired that the second derivative $\tau''(x)$ be bounded within the whole interval of interest. For cubic splines, this derivative has a piecewise linear distribution; hence, in order to have it bounded in $[0,\ 2]$, it is sufficient to assign its values at the supporting points $x_1\ (=0),\ x_2,\ \ldots,\ x_n\ (=2)$. These values will be denoted by τ_i'', for $i = 1, \ldots, n$. Furthermore, it will be assumed that $2m+1$ equally-spaced supporting points are defined in $0 \le x \le 1$. If the interval $1 \le x \le 2$ is divided into subintervals of the same size as those defined previously, then a total of $n = 4m+1$ supporting points will be defined in $0 \le x \le 2$. Clearly, due to the imposed symmetries, τ_i'' has to be specified only in the interval $0 \le x_i \le 1$, $i = 1, \ldots, 2m+1$. The remaining values of τ_i'' are simply generated by a mirror-imaging of the first $2m+1$ values about line $x = 1.0$. We illustrate these ideas with the example below.

2.4.3 Example: A Spline-Based Smooth Motion

We assume an acceleration distribution at the supporting abscissae of the spline. Here, we have infinitely many choices, but choose one that best suits the problem at hand. For example, we will assume an harmonic distribution, namely,

$$\tau_k'' = A \sin \pi x_k, \quad x_k = \frac{k-1}{m}, \qquad k = 1, \ldots, 2m+1 \tag{2.4.7a}$$

$$\tau_{m'+k}'' = \tau_{m'-k}'', \qquad k = 1, \ldots, 2m, \quad m' \equiv 2m+1 \tag{2.4.7b}$$

where A is the amplitude of τ'', a control parameter.

Note that conditions (2.4.4) and (2.4.5) can be verified with an harmonic distribution of $\tau''(x)$ if a cycloidal motion is used, as discussed in Subsection 2.3.5. However, in order to illustrate the effectiveness of cubic splines, we use a cubic spline with a simple harmonic distribution of its second derivative.

Now, eq. (2.4.3*b*) can be written as:

$$\mathbf{A}\,\mathbf{t}'' = 6\,\mathbf{C}\,\mathbf{t} \tag{2.4.8a}$$

where $\mathbf{A}$ and $\mathbf{C}$ are now $(n-2) \times (n-2)$ matrices and $\mathbf{t}$ and $\mathbf{t}''$ are the $(n-2)$-dimensional vectors defined below

$$\mathbf{t} = [\tau_2, \ldots, \tau_{n-1}]^T, \quad \mathbf{t}'' = [\tau_2'', \ldots, \tau_{n-1}'']^T \tag{2.4.8b}$$

In fact, from the assumed periodicity of $\tau(x)$, one has

$$\tau_1 = \tau_n = 0, \qquad \tau_1' = \tau_n' = 0 \tag{2.4.8c}$$

whereas, from eqs. (2.4.7*a* & *b*),

$$\tau_1'' = \tau_n'' = 0 \tag{2.4.8d}$$

the foregoing zero values thus not appearing in vectors $\mathbf{t}$ and $\mathbf{t}''$. Moreover, since the supporting points of the spline are equally spaced, the subintervals defined by two successive supporting points are of length $\Delta x_k = 1/2m$, for all k, and hence, the $(n-2) \times (n-2)$ $\mathbf{A}$ and $\mathbf{C}$ matrices take on the forms given below:

$$\mathbf{A} = \frac{1}{2m} \begin{bmatrix} 4 & 1 & 0 & 0 & \cdots & 0 \\ 1 & 4 & 1 & 0 & \cdots & 0 \\ 0 & 1 & 4 & 1 & \cdots & 0 \\ \vdots & \vdots & \ddots & \ddots & \ddots & \vdots \\ 0 & 0 & \cdots & 1 & 4 & 1 \\ 0 & 0 & 0 & \cdots & 1 & 4 \end{bmatrix} \tag{2.4.8e}$$

$$\mathbf{C} = 2m \begin{bmatrix} -2 & 1 & 0 & 0 & \cdots & 0 \\ 1 & -2 & 1 & 0 & \cdots & 0 \\ 0 & 1 & -2 & 1 & \cdots & 0 \\ \vdots & \vdots & \ddots & \ddots & \ddots & \vdots \\ 0 & 0 & \cdots & 1 & -2 & 1 \\ 0 & 0 & 0 & \cdots & 1 & -2 \end{bmatrix} \tag{2.4.8f}$$

Moreover, the $\{\tau_k\}_2^{4m-1}$ values can be readily computed from vector $\mathbf{t}$, that is, calculated in turn as

$$\mathbf{t} = \frac{1}{6}\mathbf{C}^{-1}\mathbf{A}\,\mathbf{t}'' \tag{2.4.9}$$

which requires the inversion of matrix $\mathbf{C}$. Note that matrix $\mathbf{C}$ is not only invertible, but also negative-definite and tridiagonal. Moreover, since its non-zero elements follow a simple pattern, its inverse is simple as well and can be expressed as

$$\mathbf{C}^{-1} = \frac{1}{8m^2} \begin{bmatrix} 4m-1 & 4m-2 & 4m-3 & \cdots & 2 & 1 \\ 4m-2 & 2(4m-2) & 3(4m-2) & \cdots & 4 & 2 \\ \vdots & \vdots & \vdots & \ddots & \vdots & \vdots \\ 2 & 4 & 6 & \cdots & 2(4m-2) & 4m-2 \\ 1 & 2 & 3 & \cdots & 4m-2 & 4m-1 \end{bmatrix} \tag{2.4.10}$$

Next, the $\mathbf{C}^{-1}\mathbf{A}$ product is expanded as

$$\mathbf{C}^{-1}\mathbf{A} = \frac{1}{16m^3}(\mathbf{L} + \mathbf{D} + \mathbf{U}) \tag{2.4.11}$$

where $\mathbf{L}$ is *strictly* lower triangular, $\mathbf{D}$ is diagonal, and $\mathbf{U}$ is *strictly* upper triangular. The adjective "strictly" refers here to a matrix whose diagonal has zero entries. The non-zero elements of $\mathbf{L}$, $\mathbf{D}$, and $\mathbf{U}$ are denoted by l_{ij}, d_{ii}, and u_{ij}, respectively. These are given below as

$$l_{ij} = 6\,(4m - i)\,j, \quad i > j \tag{2.4.12a}$$

$$d_{ii} = 6\,(4m - i) - 4m \tag{2.4.12b}$$

$$u_{ij} = 6\,(4m - j)\,i, \quad i < j \tag{2.4.12c}$$

for $i,\, j = 1, 2, \ldots, 4m - 1$.

On the other hand, given the symmetry of $\{\tau_k''\}_2^{4m}$, one has

$$\tau_{4m+2-k}'' = \tau_k'', \quad \text{for} \quad k = 1, \ldots, 2m+1$$

We now derive the general expression of τ_{k+1}, for $1 \leq k \leq 2m$, the corresponding expression for $2m+1 \leq k \leq 4m+1$ having been obtained by symmetry. Thus, considering such symmetries, one obtains,

$$\begin{aligned}\tau_{k+1} = \frac{1}{16m^3}\,[\,&(l_{k+1,1} + u_{k+1,4m-1})\,\tau_2'' + (l_{k+1,2} + u_{k+1,4m-2})\,\tau_3'' + \ldots \\ &+ (l_{k+1,k} + u_{k+1,4m-k})\,\tau_{k+1}'' + (d_{k+1,k+1} + u_{k+1,4m-k-1})\,\tau_{k+2}'' + \ldots \\ &+ (u_{k+1,2m-2} + u_{k+1,2m+2})\,\tau_{2m-1}'' \\ &+ (u_{k+1,2m-1} + u_{k+1,2m+1})\,\tau_{2m}''\,]\end{aligned} \tag{2.4.13}$$

an expression that can be reduced to

$$\begin{aligned}\tau_{k+1} = \frac{1}{16m^3}[&24m\tau_2'' + 48m\tau_3'' + \ldots + 24mk\tau_{k+1}'' + 24m(k+1)\tau_{k+2}'' \\ &+ \ldots + 24m(k+1)\tau_{2m-1}'' + 24m(k+1)\tau_{2m}'']\end{aligned} \tag{2.4.14a}$$

and hence,

$$\tau_{k+1} = \frac{3}{2m^2}\left[\sum_{j=1}^{k} j\tau_{j+1}'' + (k+1)\sum_{j=k+1}^{2m-1} \tau_{j+1}''\right]$$

Now, substituting the given values of $\{\tau_k''\}_2^{2m-1}$ into the above expression, one has

$$\tau_{k+1} = \frac{3A}{2m^2}\left[\sum_{j=1}^{k} j \sin\frac{j\pi}{m} + (k+1)\sum_{j=k+1}^{2m-1} \sin\frac{j\pi}{m}\right] \tag{2.4.14b}$$

The summations of eq.(2.4.14b) can be computed symbolically. Indeed, using an exponential representation of the sine function, one can derive

$$\sum_{j=k+1}^{2m} \sin\frac{j\pi}{m} = \sum_{j=k+1}^{2m} \frac{1}{2}\left[e^{i(j\pi/m)} - e^{-i(j\pi/m)}\right] =$$

$$= -\frac{\sin\frac{k\pi}{2m}\sin\frac{(k+1)\pi}{2m}}{\sin\frac{\pi}{2m}} \tag{2.4.15a}$$

where i is the imaginary unity, i.e., $i \equiv \sqrt{-1}$. Likewise,

$$\sum_{j=1}^{k} j\sin\frac{j\pi}{m} = k\frac{\sin\frac{k\pi}{2m}\sin\frac{(k+1)\pi}{2m}}{\sin\frac{\pi}{2m}} - \frac{k}{2\tan\frac{\pi}{2m}} + \frac{\sin\frac{k\pi}{m}}{4\sin^2\frac{\pi}{2m}} \tag{2.4.15b}$$

Hence,

$$\sum_{j=1}^{2m-1} j\sin\frac{j\pi}{m} \equiv \sum_{j=1}^{2m} j\sin\frac{j\pi}{m} = -\frac{m}{\tan\frac{\pi}{2m}} \tag{2.4.15c}$$

Substituting the last identities into eq. (2.4.14b), one obtains the expression below:

$$\tau_{k+1} = \frac{A}{48m^2}\left[6k\cot\frac{\pi}{2m} + (2\sin^2\frac{\pi}{2m} - 3)\frac{\sin\frac{k\pi}{m}}{\sin^2\frac{\pi}{2m}}\right] \tag{2.4.16}$$

Once the values of τ_k and τ_k'', for $k = 1, \ldots, 4m+1$, are known, those of τ_k' are readily computed as (Späth, 1978):

$$\tau_k' = \frac{\Delta\tau_k}{\Delta x_k} - \frac{1}{6}\Delta x_k(\tau_{k+1}'' + 2\tau_k''), \quad k = 1, \ldots, 4m \tag{2.4.17}$$

where Δx_k and $\Delta\tau_k$ are defined as

$$\Delta x_k = x_{k+1} - x_k, \qquad \Delta\tau_k = \tau_{k+1} - \tau_k$$

Now, substituting τ_k and τ_k'' into eq.(2.4.17), one obtains

$$\tau_{k+1}' = \frac{A}{4m}\left[\cot\frac{\pi}{2m} - \sin\frac{k\pi}{m} - \frac{\cos\frac{(2k+1)\pi}{2m}}{\sin\frac{\pi}{2m}}\right] \tag{2.4.18}$$

To derive the normal motion, all we need is to select A so that $\tau_{2m+1} = 1$, namely,

$$A = 4m\tan\frac{\pi}{2m} \tag{2.4.19}$$

from which we can derive an interesting result, i.e.,

$$\lim_{m \to \infty} A = 2\pi \tag{2.4.20}$$

Now, substituting the value of A given by eq. (2.4.19) into eqs. (2.4.16), (2.4.18), and (2.4.7b), one obtains the expressions for τ_{k+1} and τ'_{k+1}, namely,

$$\tau_{k+1} = \frac{k}{2m} + \left[2\sin^2\frac{\pi}{2m} - 3\right]\frac{\sin\frac{k\pi}{m}}{6m\sin\frac{\pi}{m}} \tag{2.4.21}$$

$$\tau'_{k+1} = 1 - \tan\frac{\pi}{2m}\sin\frac{k\pi}{m} - \frac{\cos\frac{(2k+1)\pi}{2m}}{\cos\frac{\pi}{2m}} \tag{2.4.22}$$

From eq. (2.4.22), it can be readily shown that

$$\tau'_{m+1} = 2, \quad \text{for} \quad m \geq 2 \tag{2.4.23}$$

i.e., the maximum value of the spline slope is 2, *regardless of the value of* m. Moreover,

$$\tau'_1 = \tau'_{2m+1} = 0$$

thereby satisfying the end conditions, i.e., eqs. (2.4.5a & b).

Now, finding the coefficients of the normal motion in terms of m is straightforward. In fact, substituting the values of τ_k and τ''_k into eqs.(2.4.2a–e), one obtains expressions for those coefficients, namely,

$$A_k = \frac{4}{3}m^2\tan\frac{\pi}{2m}\left[\sin\frac{k\pi}{m} - \sin\frac{(k-1)\pi}{m}\right] \tag{2.4.24a}$$

$$B_k = 2m\tan\frac{\pi}{2m}\sin\frac{(k-1)\pi}{m} \tag{2.4.24b}$$

$$C_k = 1 + \frac{2\sin^2\frac{\pi}{2m} - 3}{3\sin\frac{\pi}{m}}\left[\sin\frac{k\pi}{m} - \sin\frac{(k-1)\pi}{m}\right] - \frac{1}{3}\tan\frac{\pi}{2m}\left[\sin\frac{k\pi}{m} + 2\sin\frac{(k-1)\pi}{m}\right] \tag{2.4.24c}$$

$$D_k = \frac{k-1}{2m} + \frac{2\sin^2\frac{\pi}{2m} - 3}{6m\sin\frac{\pi}{m}}\sin\frac{(k-1)\pi}{m} \tag{2.4.24d}$$

With the aid of the foregoing results and the aforementioned acceleration distribution, we synthesized numerically several normalized curves. A list of values of m (≥ 2)—the minimum value that m can attain is 2 for nontrivial solutions—and the corresponding values of the amplitude A, which will produce a unit maximum value of τ, is given in Table 2.4.1. It should be pointed out that the maximum slope for all the synthesized curves is 2.0, *regardless of the number of supporting points*. From Table 2.4.1, it is apparent that A tends to 2π when $m \to \infty$, as predicted by eq. (2.4.20). A spline was synthesized with $m = 10$ and $A = 6.335377612981697$. Plots of the normal motion, $\tau(x)$, and its first three derivatives are shown in Figs. 2.4.1–2.4.4. In the limit, as the number of supporting points approaches ∞, $k/2m$ approaches x, and hence, from eq. (2.4.21),

$$\lim_{m\to\infty} \tau_{k+1} = x + \lim_{m\to\infty} \frac{1}{3m} \frac{\sin^2 \frac{\pi}{2m}}{\sin \frac{\pi}{m}} - \lim_{m\to\infty} \frac{1}{2m} \frac{\sin 2\pi x}{\sin \frac{\pi}{m}}$$

Table 2.4.1. Values of m and A

m	A
2	8.000000000000018
3	6.928203230275543
4	6.627416997969570
5	6.498393924658190
6	6.430780618347038
7	6.390817282924330
8	6.365195756149229
9	6.347771305504950
10	6.335377612981697
20	6.326244935779100
30	6.319319884195322
40	6.313943127557925
50	6.309684634406659
60	6.306254115941095
70	6.303449814859090
80	6.301127944458677
90	6.299183773867269
100	6.297539546259094
1000	6.283189999999999

We calculate each of these limits below:

$$\lim_{m\to\infty} \frac{1}{3m} \frac{\sin^2 \frac{\pi}{2m}}{\sin \frac{\pi}{m}} \equiv \lim_{m\to\infty} \frac{1}{3m} \frac{\sin^2 \frac{\pi}{2m}}{2 \sin \frac{\pi}{2m} \cos \frac{\pi}{2m}} =$$

$$= \lim_{m\to\infty} \frac{1}{6m} \tan \frac{\pi}{m} = 0$$

In order to calculate the last limit, let μ be the reciprocal of m, i.e.,

$$\lim_{m\to\infty} \frac{1}{2m} \frac{\sin 2\pi x}{\sin \frac{\pi}{m}} = \sin 2\pi x \lim_{\mu\to 0} \frac{\mu}{2 \sin \pi\mu}$$

$$= \sin 2\pi x \lim_{\mu\to 0} \frac{1}{2 \frac{\sin \pi\mu}{\mu}}$$

i.e.,

$$\lim_{m\to\infty} \frac{1}{2m} \frac{\sin 2\pi x}{\sin \frac{\pi}{m}} = \frac{1}{2\pi} \sin 2\pi x$$

and hence,

$$\lim_{m\to\infty} \tau_{k+1} = x - \frac{1}{2\pi} \sin 2\pi x$$

which is the cycloidal motion, eq. (2.3.6a), as expected.

Although the acceleration program prescribed in the foregoing example was so simple that its displacement program could have been derived by simple quadrature, it illustrates the effectiveness of splines as a useful tool. Additionally, it illustrates how a number of end conditions beyond those allowed by cubic splines, can be met by introducing suitable symmetries. Note also that, regardless of how complicated the acceleration program is, the displacement program can be computed efficiently and reliably with the aid of cubic periodic splines.

Once $\tau(x)$ is obtained as outlined above, the next step is the synthesis of the displacement program. This step is discussed in Section 2.5.

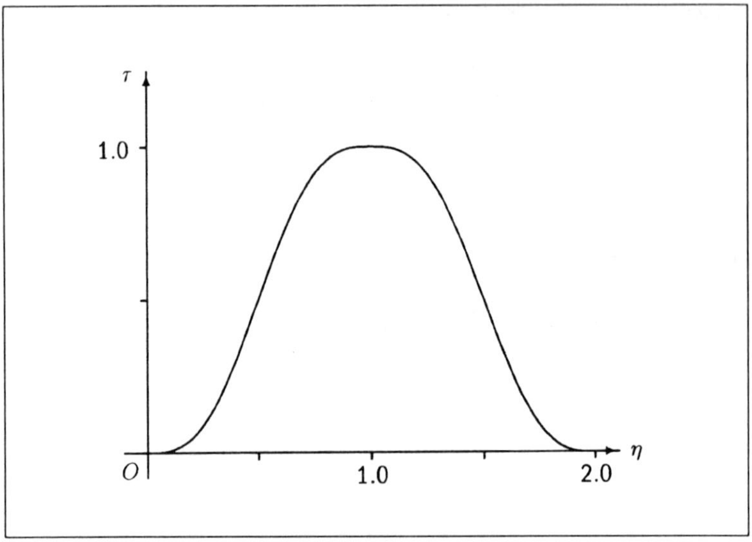

Fig. 2.4.1. The spline-synthesized displacement program

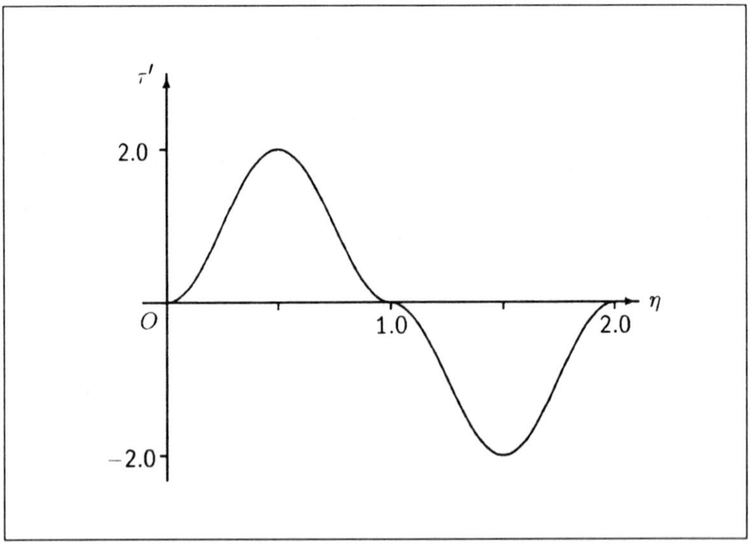

Fig. 2.4.2. First derivative of the spline-synthesized displacement program

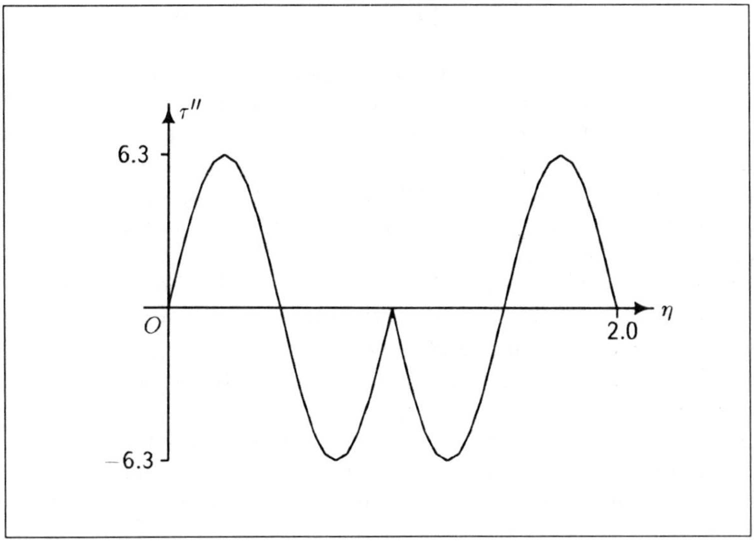

Fig. 2.4.3. Second derivative of the spline-synthesized displacement program

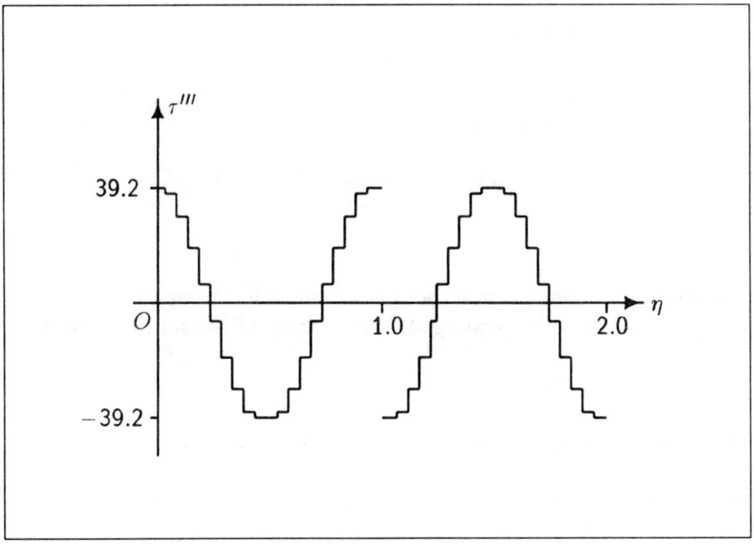

Fig. 2.4.4. Third derivative of the spline-synthesized displacement program

2.5 Synthesis of the Displacement Program

In this section, the synthesis of a follower-displacement program, using the normal motion described in Section 2.2, is presented. Let the displacement of the follower be given by $c + \sigma(\psi)$, where c is a constant representing the minimum value of that displacement, while $\sigma(\psi)$ was defined in Section 2.2. Moreover, the maximum value of $\sigma(\psi)$ is h, the rise of the follower. Clearly, c, h, $\sigma(\psi)$, and s will have units of length for the case of translating followers; for oscillating followers, $s(\psi)$ is replaced by $\phi(\psi)$, which has units of radians. Constant c is in turn computed from considerations concerning the profile synthesis, as will be shown in Chapters 4, 5, and 6, since the objective of this subsection is the synthesis of $\sigma(\psi)$.

The displacement program for the follower of a cam mechanism is now assumed to have a lower dwell in the interval $[0,\ \psi_a]$; then, a rise in the interval $[\psi_a,\ \psi_b]$; next, a higher dwell in the interval $[\psi_b,\ \psi_c]$; and, finally, a return in the interval $[\psi_c,\ 2\pi]$. For the rise interval, function $\sigma(\psi)$ can be obtained by properly scaling and translating the first half of the normal motion. For the return interval, similar operations are done on the second half of the normal motion. That is, function $\sigma(\psi)$ is obtained as:

$$\sigma(\psi) = 0, \qquad 0 \le \psi \le \psi_a \tag{2.5.1a}$$

$$\sigma(\psi) = h\,\tau(x), \qquad x = \frac{\psi - \psi_a}{\psi_b - \psi_a}, \qquad \psi_a \le \psi \le \psi_b \tag{2.5.1b}$$

$$\sigma(\psi) = h, \qquad \psi_b \le \psi \le \psi_c \tag{2.5.1c}$$

$$\sigma(\psi) = h\tau(1 - x), \qquad x = \frac{\psi - \psi_c}{2\pi - \psi_c}, \qquad \psi_c \le \psi \le 2\pi \tag{2.5.1d}$$

In summary, once the normal motion is available, the synthesis of the rise and the return phases reduces to scaling and translating of the normal motion.

2.5.1 Example: A Spline-Based *DRDR* Motion Program

Here we want to synthesize a *DRDR* motion program for an oscillating follower. Moreover, the follower is to sweep an angle of 60° with a lower dwell taking 25% and an upper dwell taking 10% of the whole cycle. Additionally, the rise phase is to take place in 50 % of the cycle, the return phase taking place in the remaining 15%.

We use the motion synthesized in Example 2.4.3 with $m = 10$. We then have, for σ and ψ in radians,

$$\sigma(\psi) = 0, \qquad 0 \le \psi \le \frac{\pi}{2}$$

$$\sigma(\psi) = \frac{\pi}{3}\tau(x), \qquad x = \frac{\psi}{\pi} - \frac{1}{2}$$

$$\sigma(\psi) = \frac{\pi}{3}, \qquad \frac{3\pi}{2} \le \psi \le \frac{17\pi}{10}$$

$$\sigma(\psi) = \frac{\pi}{3}\tau(1-x), \qquad x = \frac{10\psi}{3\pi} - \frac{17}{3}$$

The motion program thus produced, along with its first and second derivatives, is shown in Fig. 2.5.1, 2.5.2*a*) and *b*), respectively.

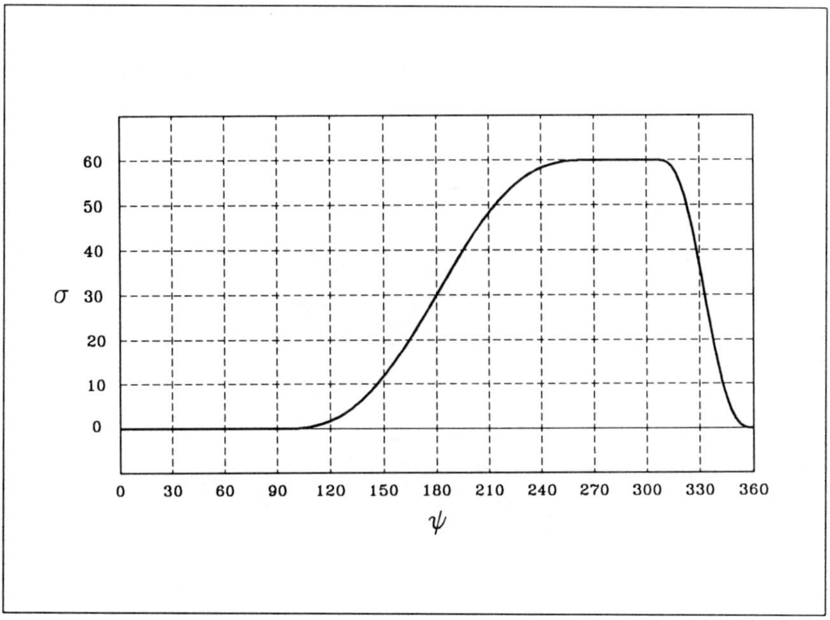

Fig. 2.5.1. Motion program for Example 2.5.1: displacement.

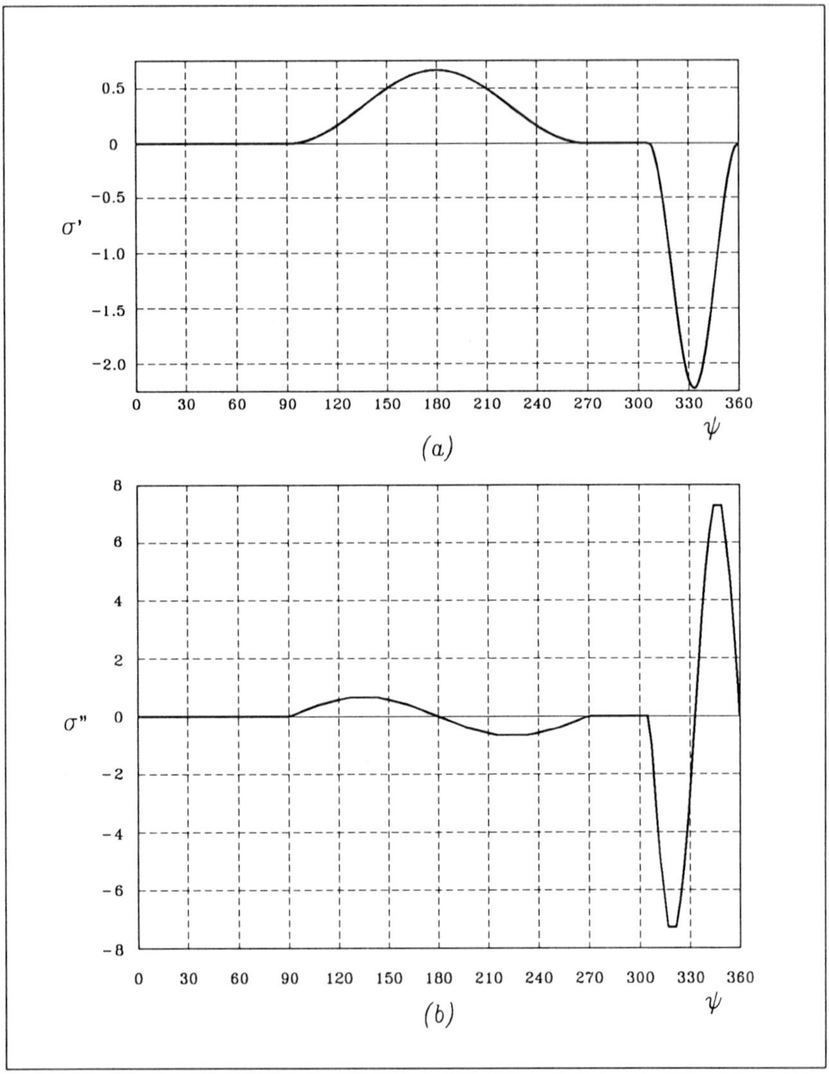

Fig. 2.5.2. Motion program for Example 2.5.1: *a*) first derivative, and *b*) second derivative.

Chapter 3 CAM PROFILE GEOMETRY

3.1 Introduction

In Chapters 4 and 5, the generation of the cam profile of disk cams will be discussed. That is, the optimum-design algorithms studied in those chapters will yield a discrete set of contour points $\{P_i\}_1^m$ with Cartesian coordinates x_i and y_i for $i = 1, \ldots, m$. For purposes of rendering the cam profile either on a monitor or on a metal plate using an NC machine tool, a database with a set $\mathcal{S}$ of $\{(x_i,\ y_i)\}_1^m$ pairs of point coordinates will suffice. Obviously, the larger the set, the smoother the rendering. However, m cannot grow without limit because *i)* beyond a certain value of m dictated by the resolution of the rendering medium, no apparent improvement is reached by increasing m, and *ii)* due to physical limitations, databases are of a finite size. Hence, a reasonable value of m must be selected a priori.

Once a suitable number of cam points is available, one must cope with other problems in order to produce accurate values of both *local* and *global* geometric properties of the contour. The former involve slope and curvature values throughout the contour; the latter, volumetric properties, i.e., area or, correspondingly, volume, centroid location, and moments and products of inertia of the cam.

Thus, global geometric properties involve the calculation of integrals, which comprise numerically stable computations. That is, truncation errors induced by the discretization of the contour are attenuated due to the *filtering* effect of integrations, although care must be taken in order to produce results that are accurate enough. Various techniques are currently available for the systematic evaluation of the integrals that arise, those yielding the highest accuracy for a fixed value of m being techniques based on a spline representation of the contour. These will be discussed in Chapter 8.

However, regarding the calculation of local properties, the truncation errors due to the contour discretization are amplified by virtue of the differentiations involved. Here, again, various techniques are available to cope with these problems; the techniques favored in this book are those based on a spline representation of the contour.

Thus, a methodology for handling the discrete points via periodic parametric cubic splines will be discussed in Chapter 8. In Section 3.2 various concepts of differential geometry are discussed; these involve local properties such as slope, which is needed in the calculation of the pressure angle, and cam curvature, which arises in avoiding cusps and *undercutting*. In Section 3.3, the global properties of the cam such as area, volume, centroid location, and moments of inertia are described.

3.2 Local Properties

In this section, the differential-geometry relationships pertaining to cam profiles are discussed. These relationships will be used later for determining the cam profile and some related variables such as the tangent orientation, required to evaluate the pressure angle, and the curvature of the cam profile, required to verify the appearance of cusps or undercutting. This information is also needed in planning the trajectory of the cutter of an NC machine tool, as outlined in Chapter 9.

3.2.1 Tangent at a Point of a Cam Profile

The relationship for determining the angle between the position vector of a point on the cam profile and the tangent at that point is derived first. This result will be used for the derivation of the synthesis equations of the cam profile as shown in Chapters 4 and 5.

Reference is made now to Fig. 3.2.1. In that figure, λ measures the arc length from the origin O_λ, while $\mathbf{e}_t$ is a unit vector parallel to line T, the tangent at point Q of the curve Γ. Note that $\mathbf{e}_t$ points in the direction in which λ grows.

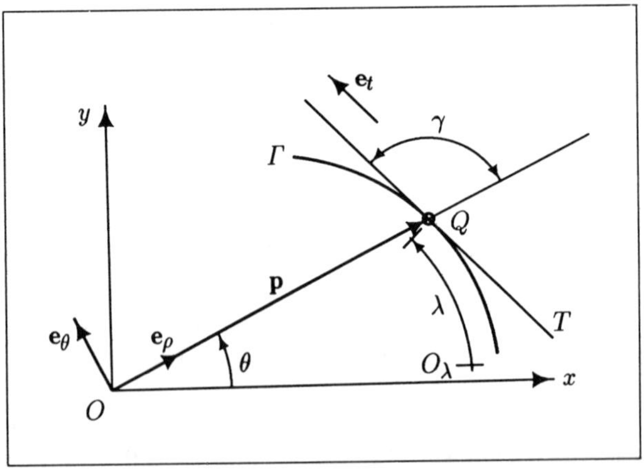

Fig. 3.2.1. Geometry of a planar curve.

What we need to derive as an expression for vector $\mathbf{e}_t$ is a parametric representation of the position vector $\mathbf{p}$ of an arbitrary point of the contour, of components x and y, i.e., $\mathbf{p} = [x,\ y]^T$. However, x and y are usually available as functions of a parameter q different from λ. Hence, derivatives with respect to λ will be obtained using the *chain rule*. First, from elementary differential geometry (Hurley, 1981), we have

$$\mathbf{e}_t = \frac{d\mathbf{p}}{d\lambda} = \frac{d\mathbf{p}}{dq}\frac{dq}{d\lambda} = \frac{\mathbf{p}'(q)}{\lambda'(q)} \tag{3.2.1a}$$

where

$$\mathbf{p}'(q) = \begin{bmatrix} x'(q) \\ y'(q) \end{bmatrix} \tag{3.2.1b}$$

Now, since $\mathbf{e}_t$ is a unit vector, we have

$$\lambda'(q) = \pm\|\mathbf{p}'(q)\| = \pm\sqrt{x'^2(q) + y'^2(q)} \tag{3.2.2a}$$

where, obviously, the negative sign is used if $\lambda'(q) < 0$. Thus,

$$\mathbf{e}_t = \begin{cases} \dfrac{\mathbf{p}'(q)}{\|\mathbf{p}'(q)\|}, & \text{if } \lambda'(q) > 0 \\[2ex] -\dfrac{\mathbf{p}'(q)}{\|\mathbf{p}'(q)\|}, & \text{if } \lambda'(q) < 0 \end{cases} \tag{3.2.2b}$$

Sometimes the cam contour is given in polar coordinates ρ and θ in the form $\rho = \rho(\theta)$ or, equivalently, in parametric form as $\rho = \rho(\psi)$ and $\theta = \theta(\psi)$. The position vector of a point on the cam profile will then be given in terms of two mutually orthogonal unit vectors, $\mathbf{e}_\rho$ and $\mathbf{e}_\theta$, directed along the directions in which ρ and θ are measured, respectively. Thus,

$$\mathbf{p} = \rho(\theta)\,\mathbf{e}_\rho \tag{3.2.3a}$$

Differentiating both sides of eq. (3.2.3*a*) with respect to the arc length λ yields

$$\frac{d\mathbf{p}}{d\lambda} \equiv \mathbf{e}_t = \rho(\theta)\frac{d\theta}{d\lambda}\mathbf{e}_\theta + \frac{d\rho}{d\theta}\frac{d\theta}{d\lambda}\mathbf{e}_\rho \tag{3.2.3b}$$

where the *chain rule* and the fact that $d\mathbf{e}_\rho/d\theta = \mathbf{e}_\theta$ have been used. By dot multiplying both sides of eq. (3.2.3*b*) by $\mathbf{e}_\theta$ and $\mathbf{e}_\rho$, one obtains

$$\mathbf{e}_\theta \cdot \mathbf{e}_t \equiv \sin\gamma = \rho(\theta)\frac{d\theta}{d\lambda}, \qquad \mathbf{e}_\rho \cdot \mathbf{e}_t \equiv \cos\gamma = \frac{d\rho}{d\theta}\frac{d\theta}{d\lambda} \tag{3.2.4a}$$

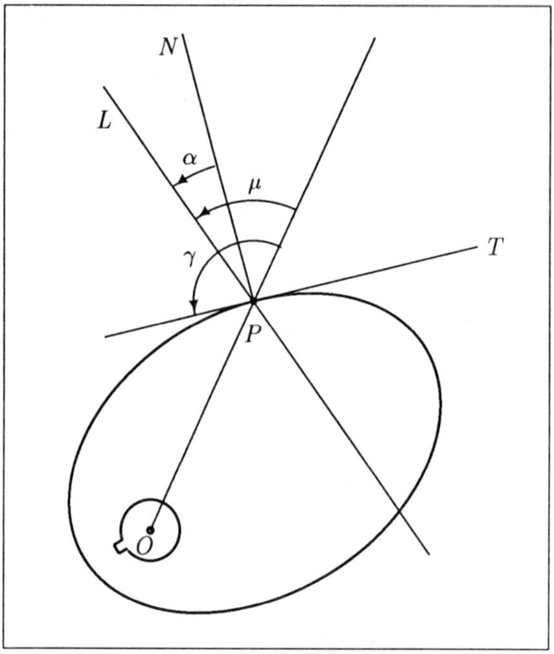

Fig. 3.2.2. Pressure angle determination.

Furthermore, combining eqs. (3.2.4a) yields

$$\tan\gamma = \frac{\rho(\theta)}{\rho'(\theta)}, \qquad \rho'(\theta) \equiv \frac{d\rho}{d\theta} \tag{3.2.4b}$$

This result will be used in Chapters 4 and 5 for deriving the synthesis equations of the cam profile.

The direction of the tangent to the cam profile, given by the angle γ between the tangent and the radius vector ρ, is used to determine the *pressure angle of the cam profile.* This angle was defined in Section 1.4.1 as the angle between the normal to the cam profile and a given direction. This direction is that of the velocity of the point of the follower at which the force exerted by the cam is applied. Let us assume in this discussion that the direction of interest is that of L in Fig. 3.2.2, the normal to the cam profile being denoted by N.

Since we assume L to be known, we know the angle between L and line OP, called μ here. With this information, we should be able to calculate the pressure angle α in terms of γ and μ. In fact, from Fig. 3.2.2, if the difference Δ between μ

and α is subtracted from γ, the angle between N and T is obtained, which is $\pi/2$, i.e.,

$$\gamma - \Delta = \pi/2 \tag{3.2.5a}$$

where

$$\Delta = \mu - \alpha \tag{3.2.5b}$$

and hence,

$$\alpha = \frac{\pi}{2} - (\gamma - \mu) \tag{3.2.5c}$$

which is the desired relation. Expressions for μ vary from case to case, but the principle is the same. In particular, μ depends on the angle of rotation of the cam, ψ, and the position of the follower, whether translational or angular.

3.2.2 Cam Curvature

The shape of a planar curve Γ depends on the rate of change of the direction of its tangent with respect to the arc length, a measure which is called the *curvature* of Γ and is designated hereafter as κ. This variable plays an important role in the design of cam mechanisms, for it is directly related to the appearance of cusps and the already mentioned effect known as *undercutting*—see Chapters 1 and 5. The reciprocal of the curvature, κ, is the radius of curvature, r, i.e.,

$$r = \frac{1}{\kappa} \tag{3.2.6}$$

The radius of curvature at a point of the cam profile is the radius of a circle tangent at that point to the cam profile, on the concave side as shown in Fig. 3.2.3. The curvature of that circle is the same as that of the cam profile. For our purposes, the radius of curvature is positive if the center K of the circle is located between the center of rotation O and the point of tangency Q; otherwise, the radius of curvature is negative. In the discussion which follows, formulas for calculating the curvature at any point of a cam profile are derived.

Let us consider a planar curve, as shown in Fig. 3.2.4. At any point P on the curve, whose position vector is denoted by $\mathbf{p}$, a unique orthonormal pair of vectors is defined, namely, the tangent and the normal vectors indicated in that figure as $\mathbf{e}_t$ and $\mathbf{e}_n$. Let λ measure the arc length as in the foregoing discussion. Unit vectors $\mathbf{e}_t$ and $\mathbf{e}_n$, their derivatives with respect to λ, and the curvature κ are related by the Frenet-Serret formulas (Jeffrey, 1989; Adams, 1987; Brand, 1965), namely,

$$\frac{d\mathbf{e}_t}{d\lambda} = \kappa\, \mathbf{e}_n \tag{3.2.7a}$$

$$\frac{d\mathbf{e}_n}{d\lambda} = -\kappa\, \mathbf{e}_t \tag{3.2.7b}$$

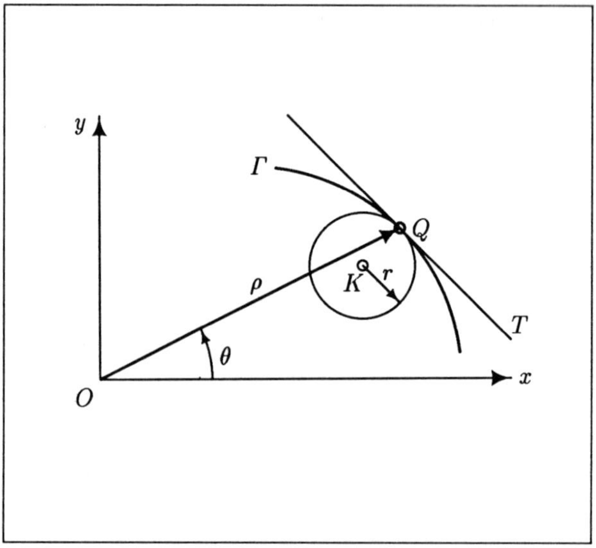

Fig. 3.2.3. Radius of curvature of a cam profile.

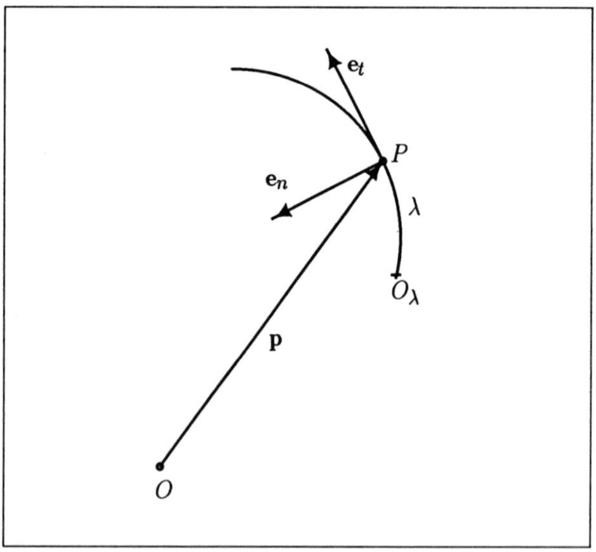

Fig. 3.2.4. Planar curve and the orthonormal vectors $\mathbf{e}_t$ and $\mathbf{e}_n$.

Now, let the position vector $\mathbf{p}$ be a function of a parameter q that, in general, is different from λ. Differentiation of $\mathbf{p}$ with respect to q twice yields

$$\mathbf{p}'(q) \equiv \frac{d\lambda}{dq}\frac{d\mathbf{p}}{d\lambda} = \lambda'(q)\,\mathbf{e}_t \tag{3.2.8a}$$

$$\mathbf{p}''(q) = \lambda''(q)\,\mathbf{e}_t + \lambda'^2(q)\,\kappa\,\mathbf{e}_n \tag{3.2.8b}$$

where eq. (3.2.7a) has been used. Performing the cross product $\mathbf{p}'(q) \times \mathbf{p}''(q)$, one has

$$\mathbf{p}'(q) \times \mathbf{p}''(q) = \lambda'^3(q)\,\kappa\,\mathbf{e}_b \tag{3.2.9a}$$

where $\mathbf{e}_b$ is the unit binormal vector of the curve, defined as

$$\mathbf{e}_b = \mathbf{e}_t \times \mathbf{e}_n \tag{3.2.9b}$$

and all vectors involved are assumed to be defined in 3-D space. However, we defined $\mathbf{p}$ above as a 2-D vector, an inconsistency that can be resolved by simply adding a third zero component to $\mathbf{p}$ and all other 2-D vectors involved.

If we now complete the planar coordinate system x, y with a third axis z defining a right-hand coordinate frame x, y, z and attach unit vectors $\mathbf{i}$, $\mathbf{j}$, $\mathbf{k}$ to the directions of these axes, then, obviously, $\mathbf{e}_b = \mathbf{k}$ at *convex* points of the cam profile, whereas $\mathbf{e}_b = -\mathbf{k}$ at *concave* points, as shown in Fig. 3.2.5, in which it is assumed that $\lambda'(q) > 0$. From this figure, it is clear that the cross product $\mathbf{p}'(q) \times \mathbf{p}''(q)$, appearing in eq. (3.2.9a), points outside of the plane of the figure at concave points and inside of that plane at convex points. Furthermore, if we recall eqs. (3.2.2b), we can rewrite eq. (3.2.9a) as

$$\mathbf{p}'(q) \times \mathbf{p}''(q) = \mathrm{sgn}\,[\,\lambda'(q)\,]\,\|\,\mathbf{p}'(q)\,\|^3\,\kappa\,\mathbf{k} \tag{3.2.10}$$

where sgn$(\cdot)$ is the *signum* function of $(\cdot)$, which is defined as $+1$ if its argument is positive; it is -1 if its argument is negative. If the argument vanishes, then we can define sgn$(\cdot)$ arbitrarily as zero.

Now we can solve for κ from eq. (3.2.10) by simply dot multiplying both sides of this equation by $\mathbf{k}$, thereby deriving the desired relation, namely,

$$\kappa = \mathrm{sgn}\,[\,\lambda'(q)\,]\,\frac{\mathbf{p}'(q) \times \mathbf{p}''(q) \cdot \mathbf{k}}{\|\,\mathbf{p}'(q)\,\|^3} \tag{3.2.11}$$

In the derivation of eq. (3.2.11), use was made of the cross product, which is defined only in three-dimensional vector spaces. However, our discussion focuses on planar cam profiles, and hence, an ad hoc derivation of that expression is in order.

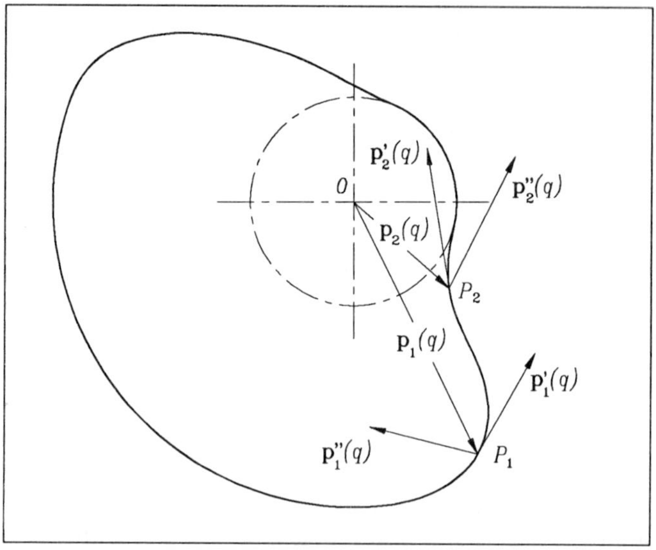

Fig. 3.2.5 Concave and convex points of a cam profile.

Below we illustrate a technique whereby we can dispense with the cross product and limit all vector calculations to the two-dimensional case. To this end, we introduce a 2×2 matrix $\mathbf{E}$ that maps any 2-D vector into a new vector rotated clockwise with respect to the former through 90° without modifying its magnitude. This matrix is defined as

$$\mathbf{E} = \begin{bmatrix} 0 & -1 \\ 1 & 0 \end{bmatrix} \tag{3.2.12a}$$

Note that $\mathbf{E}$ is an orthogonal matrix, i.e.,

$$\mathbf{E}\,\mathbf{E}^T = \mathbf{E}^T\mathbf{E} = \mathbf{1} \tag{3.2.12b}$$

where $\mathbf{1}$ denotes the 2×2 identity matrix. Moreover, $\mathbf{E}$ is skew-symmetric, i.e.,

$$\mathbf{E} = -\mathbf{E}^T \tag{3.2.12c}$$

and hence, by virtue of eq. (3.2.12*b*),

$$\mathbf{E}^2 = -\mathbf{1} \tag{3.2.12d}$$

Now, let $\mathbf{u}$ and $\mathbf{v}$ be two 3-D vectors whose components are denoted by u_i and v_i, respectively, for $i = 1,\ 2,\ 3$. The projection of $\mathbf{u} \times \mathbf{v}$ onto the z axis, denoted by p, is computed as

$$p = \mathbf{u} \times \mathbf{v} \cdot \mathbf{k}$$

Thus, p can be calculated from a determinant, namely,

$$p = \det[\mathbf{u}, \mathbf{v}, \mathbf{k}] = \det \begin{bmatrix} u_1 & v_1 & 0 \\ u_2 & v_2 & 0 \\ u_3 & v_3 & 1 \end{bmatrix} \tag{3.2.13}$$

and hence,

$$p = u_1 v_2 - u_2 v_1$$

Alternatively, p can be calculated using only 2-D vectors, which requires defining *reduced vectors* $\mathbf{u}_r$ and $\mathbf{v}_r$, as

$$\mathbf{u}_r \equiv \begin{bmatrix} u_1 \\ u_2 \end{bmatrix}, \qquad \mathbf{v}_r \equiv \begin{bmatrix} v_1 \\ v_2 \end{bmatrix}$$

and hence,

$$\mathbf{E}\,\mathbf{v}_r = \begin{bmatrix} -v_2 \\ v_1 \end{bmatrix} \tag{3.2.14a}$$

Therefore, p can be expressed as the negative of the scalar product of $\mathbf{u}_r$ and $\mathbf{E}\,\mathbf{v}_r$, namely,

$$p = -\mathbf{u}_r^T \mathbf{E}\,\mathbf{v}_r \tag{3.2.14b}$$

Moreover, since $\mathbf{E}$ is skew-symmetric, p can be written alternatively as

$$p = -\mathbf{u}_r^T \mathbf{E}^T \mathbf{v}_r \equiv \mathbf{v}_r^T \mathbf{E}\,\mathbf{u}_r \tag{3.2.14c}$$

Therefore, if the position vector $\mathbf{p}$ of point P of Fig. 3.2.3 is defined as a 2-D vector that is a function of a parameter q, i.e.,

$$\mathbf{p} = \begin{bmatrix} x(q) \\ y(q) \end{bmatrix} \tag{3.2.15}$$

then κ, as given by eq. (3.2.10) can be rewritten as

$$\kappa = -\mathrm{sgn}[\lambda'(q)]\frac{\mathbf{p}'^T(q)\,\mathbf{E}\,\mathbf{p}''(q)}{\|\mathbf{p}'(q)\|^3} \equiv \mathrm{sgn}[\lambda'(q)]\frac{\mathbf{p}''^T(q)\,\mathbf{E}\,\mathbf{p}'(q)}{\|\mathbf{p}'(q)\|^3} \tag{3.2.16}$$

Furthermore, if $\cos(\mathbf{p}'', \mathbf{E}\,\mathbf{p}')$ denotes the cosine of the angle between vectors $\mathbf{p}'$ and $\mathbf{p}''$, expression (3.2.16) is reduced to

$$\kappa = \operatorname{sgn}[\lambda'(q)] \frac{||\mathbf{p}'||\,||\mathbf{p}''||\cos(\mathbf{p}'', \mathbf{E}\,\mathbf{p}')}{||\mathbf{p}'||^3} =$$

$$= \operatorname{sgn}[\lambda'(q)] \frac{||\mathbf{p}''||\,\cos(\mathbf{p}'', \mathbf{E}\,\mathbf{p}')}{||\mathbf{p}'||^2} \tag{3.2.17}$$

Equation (3.2.16) is a general formula for computing the curvature of any curve. In particular, if the parametric representation of a cam profile, $x = x(q)$ and $y = y(q)$, is given in terms of *any* parameter q, then the position vector, $\mathbf{p}$, can be expressed as:

$$\mathbf{p}(q) = x(q)\,\mathbf{i} + y(q)\,\mathbf{j} \tag{3.2.18}$$

where $\mathbf{i}$ and $\mathbf{j}$ are constant unit vectors defined above. Differentiation of both sides of eq. (3.2.18) with respect to q twice yields

$$\mathbf{p}'(q) = x'(q)\,\mathbf{i} + y'(q)\,\mathbf{j} \tag{3.2.19}$$

$$\mathbf{p}''(q) = x''(q)\,\mathbf{i} + y''(q)\,\mathbf{j} \tag{3.2.20}$$

Applying either eq. (3.2.10) or eqs. (3.2.17*a* or *b*), we compute the curvature as:

$$\kappa = \operatorname{sgn}[\lambda'(q)] \frac{x'y'' - y'x''}{(x'^2 + y'^2)^{3/2}} \tag{3.2.21}$$

where the argument q has been dropped for brevity.

If the cam profile is given instead in polar coordinates $\rho = \rho(\theta)$, then θ can be considered as a parameter, and the position vector of any point of the curve can be expressed as

$$\mathbf{p}(\theta) = \rho(\theta)\,\mathbf{e}_\rho \tag{3.2.22}$$

where $\mathbf{e}_\rho$ and $\mathbf{e}_\theta$ are shown in Fig. 3.2.1. After differentiating both sides of eq. (3.2.22) with respect to θ twice, one has

$$\mathbf{p}'(\theta) = \rho(\theta)\,\mathbf{e}_\theta + \rho'(\theta)\,\mathbf{e}_\rho \tag{3.2.23a}$$

$$\mathbf{p}''(\theta) = [\rho''(\theta) - \rho(\theta)]\,\mathbf{e}_\rho + 2\rho'(\theta)\,\mathbf{e}_\theta \tag{3.2.23b}$$

Upon substitution of eqs. (3.2.23*a* & *b*) into eq. (3.2.10), one obtains

$$\kappa = \operatorname{sgn}[\lambda'(q)] \frac{\rho^2 + 2\rho'^2 - \rho\rho''}{(\rho^2 + \rho'^2)^{3/2}} \tag{3.2.24}$$

where the argument θ has been dropped for compactness.

3.3 Global Properties

The global geometric properties of a planar contour are its area, the position vector of its centroid, and its moments and products of inertia about a certain point, e.g., its centroid. Likewise, the global geometric properties of a closed region of the 3-D space are its volume, the position vector of its centroid, and its 3×3 inertia matrix about a certain point such as its centroid. Below we recall the formal definitions of these items.

3.3.1 Planar Contours

Let $\mathcal{R}$ denote the region bounded by a closed contour $\mathcal{C}$ in the x-y plane, and $\mathbf{p}$ the position vector of a point P of $\mathcal{R}$ in a frame $\mathcal{F}(O,\ x,\ y,\ z)$. Moreover, its area is denoted by A, its *first moment* by vector $\mathbf{q}$, and its inertia matrix about the given origin O by $\mathbf{I}_O$. Thus,

$$A = \int_{\mathcal{R}} dA \tag{3.3.1a}$$

$$\mathbf{q} = \int_{\mathcal{R}} \mathbf{p}\, dA \tag{3.3.1b}$$

$$\mathbf{I}_O = \int_{\mathcal{R}} (\|\mathbf{p}\|^2 \mathbf{1} - \mathbf{p}\mathbf{p}^T)\, dA \tag{3.3.1c}$$

where $dA = dx\, dy$, $\mathbf{1}$ is the 2×2 identity matrix, $\|\mathbf{p}\|^2$ denotes the square of the magnitude of $\mathbf{p}$, i.e.,

$$\|\mathbf{p}\|^2 = x^2 + y^2 \tag{3.3.2a}$$

and $\mathbf{p}\mathbf{p}^T$ is the 2×2 matrix indicated below

$$\mathbf{p}\mathbf{p}^T = \begin{bmatrix} x \\ y \end{bmatrix} [x \quad y] = \begin{bmatrix} x^2 & xy \\ xy & y^2 \end{bmatrix} \tag{3.3.2b}$$

Moreover, from the *mean-value theorem* of the integral calculus (Hurley, 1981), a vector $\mathbf{c}$ exists such that

$$\mathbf{q} = A\,\mathbf{c} \tag{3.3.3}$$

Vector $\mathbf{c}$ thus defines the position of the *centroid* C of $\mathcal{R}$. Henceforth, the two components of $\mathbf{c}$ in $\mathcal{F}$ are denoted by c_1 and c_2, i.e., $\mathbf{c} = [c_1,\ c_2]^T$.

The moment of inertia of $\mathcal{R}$ about C is then defined as

$$\mathbf{I}_C \equiv \int_{\mathcal{R}} [\,\|\mathbf{p} - \mathbf{c}\|^2 \mathbf{1} - (\mathbf{p} - \mathbf{c})(\mathbf{p} - \mathbf{c})^T\,]\, dA \tag{3.3.4}$$

Moreover, $\mathbf{I}_C$ reduces to

$$\mathbf{I}_C = \begin{bmatrix} I_{xx} & I_{xy} \\ I_{xy} & I_{yy} \end{bmatrix} \tag{3.3.5a}$$

where

$$I_{xx} \equiv \int_{\mathcal{R}} (y - c_2)^2 dx\, dy \tag{3.3.5b}$$

$$I_{yy} \equiv \int_{\mathcal{R}} (x - c_1)^2 dx\, dy \tag{3.3.5c}$$

$$I_{xy} \equiv -\int_{\mathcal{R}} (x - c_1)(y - c_2)\, dx\, dy \tag{3.3.5d}$$

The quantities I_{xx} and I_{yy} are usually referred to as the centroid moments of inertia of $\mathcal{R}$ about the x and y axes, respectively, whereas I_{xy} is termed the product of inertia of $\mathcal{R}$ about its centroid. Moreover, the quantity I_{zz}, defined as

$$I_{zz} = I_{xx} + I_{yy} \tag{3.3.5e}$$

is known as the *polar* moment of inertia of $\mathcal{R}$ about its centroid.

The integral appearing in eq. (3.3.1b) is called the *first moment* of $\mathcal{R}$, whereas $\mathbf{I}_C$ is referred to as the *second moment* of $\mathcal{R}$. By extension, the integral of eq. (3.3.1a) is also termed the *zeroth moment* of $\mathcal{R}$.

Calculating the three moments of $\mathcal{R}$ directly as given above calls for a double integration, one in x and one in y. This is not practical, the foregoing integrals being more readily calculated if suitable transformation formulas are introduced, as indicated in Chapter 8, thereby reducing those calculations to line integrals. Furthermore, if the cam profile is described in polar coordinates as $\rho = \rho(\theta)$, then the area is given by

$$A = \frac{1}{2}\int_0^{2\pi} \rho^2(\theta)\, d\theta \tag{3.3.6}$$

3.3.2 Solid Regions

Although the global geometric properties of cam disks can be described as those of a planar contour, the presence of non-symmetric hubs, keyways, and the like render the cam plate a 3-D solid of a general shape. By the latter we mean a solid that cannot be generated by a simple *extrusion* or *sweeping* of the planar contour in a direction perpendicular to its plane. In this case, the global properties are those of a 3-D region $\mathcal{R}$ bounded by a closed surface $\mathcal{S}$, namely, its volume V, its vector first-moment $\mathbf{q}$, and its 3×3 inertia matrix about point O. These are defined below:

$$V \equiv \int_{\mathcal{R}} dV \tag{3.3.7a}$$

$$\mathbf{q} \equiv \int_{\mathcal{R}} \mathbf{p}\, dV \tag{3.3.7b}$$

$$\mathbf{I}_O \equiv \int_{\mathcal{R}} (\|\mathbf{p}\|^2 \mathbf{1} - \mathbf{p}\mathbf{p}^T)\, dV \tag{3.3.7c}$$

where, now, $\mathbf{1}$ is the 3×3 identity matrix, $\mathbf{p} \equiv [\, x, \; y, \; z\,]^T$ and hence,

$$\|\mathbf{p}\|^2 = x^2 + y^2 + z^2 \tag{3.3.7d}$$

while

$$\mathbf{p}\mathbf{p}^T \equiv \begin{bmatrix} x \\ y \\ z \end{bmatrix} [\, x \quad y \quad z\,] = \begin{bmatrix} x^2 & xy & xz \\ xy & y^2 & yz \\ xz & yz & z^2 \end{bmatrix} \tag{3.3.7e}$$

Again, from the mean-value theorem of the integral calculus, a vector $\mathbf{c}$ exists such that

$$\mathbf{q} = V\mathbf{c} \tag{3.3.8}$$

and vector $\mathbf{c}$, once more, defines the location of the *centroid* C of $\mathcal{R}$. Correspondingly, the three components of $\mathbf{c}$ in a frame $\mathcal{F}(O, \; x, \; y, \; z)$ are given by c_i, for $i = 1, 2, 3$, i.e., $\mathbf{c} = [\, c_1, \; c_2, \; c_3\,]^T$.

The moment of inertia of $\mathcal{R}$ about C is then defined as

$$\mathbf{I}_C \equiv \int_{\mathcal{R}} [\, \|\mathbf{p} - \mathbf{c}\|^2 \mathbf{1} - (\mathbf{p} - \mathbf{c})(\mathbf{p} - \mathbf{c})^T\,]\, dV \tag{3.3.9}$$

which then reduces to

$$\mathbf{I}_C = \begin{bmatrix} I_{xx} & I_{xy} & I_{xz} \\ I_{xy} & I_{yy} & I_{yz} \\ I_{xz} & I_{yz} & I_{zz} \end{bmatrix} \tag{3.3.10a}$$

where

$$I_{xx} \equiv \int_{\mathcal{R}} [\, (y - c_2)^2 + (z - c_3)^2\,]\, dx\, dy\, dz \tag{3.3.10b}$$

$$I_{yy} \equiv \int_{\mathcal{R}} [\,(x-c_1)^2 + (z-c_3)^2\,]\, dx\, dy\, dz \tag{3.3.10c}$$

$$I_{zz} \equiv \int_{\mathcal{R}} [\,(x-c_1)^2 + (y-c_2)^2\,]\, dx\, dy\, dz \tag{3.3.10d}$$

$$I_{xy} \equiv -\int_{\mathcal{R}} (x-c_1)\,(y-c_2)\, dx\, dy\, dz \tag{3.3.10e}$$

$$I_{xz} \equiv -\int_{\mathcal{R}} (x-c_1)\,(z-c_3)\, dx\, dy\, dz \tag{3.3.10f}$$

$$I_{yz} \equiv -\int_{\mathcal{R}} (y-c_2)\,(z-c_3)\, dx\, dy\, dz \tag{3.3.10g}$$

As in the case of planar contours, the foregoing calculations involve multiple integrals, in this instance triple. Reduced formulas transforming these calculations into surface and even line integrals are given in Chapter 8.

3.4 Evaluation of Local and Global Geometric Properties of the Cam Profile

As mentioned in Section 1.4, undercutting and cusps can occur during the cam profile synthesis. These occurrences can be verified, in turn, by computing the curvature at each point of the cam profile. Obviously, this can be done only at a discrete set of points $\{P_i\}_1^m$ of the cam profile.

Since in the design phase one has information about the displacement program, i.e., one knows $\sigma(\psi)$ rather than $\rho(\theta)$, the local geometric properties of the cam profile are computed based on $\sigma(\psi)$ and its derivatives with respect to ψ. Thus, in Chapters 4 and 5, we introduce formulas for the curvature of the cam profile in terms of the displacement program and its derivatives.

For static and dynamic balancing and also for dynamic analysis, the determination of the cam global geometric properties is very important. For example, very often the cam axis of rotation and the principal axis of inertia perpendicular to the cam disk are skew. Therefore, in order to balance cams statically and dynamically their global volumetric properties must be calculated. The calculation of these properties requires the introduction of suitable numerical techniques that are discussed in Chapter 8.

Chapter 4 OPTIMIZATION OF PLANAR CAM MECHANISMS WITH FLAT-FACE FOLLOWERS

4.1 Introduction

The problem of cam-profile determination, as pertaining to cam mechanisms with *translating* and *oscillating* flat-face followers, is handled in this chapter. The problem addressed here has been solved traditionally using graphical methods—see, e.g., (Rothbart, 1956; Wunderlich, 1971). With the introduction of the computer as an engineering design tool, computer-oriented methods have been proposed—see, e.g., (Chen, 1977) and the references contained therein. However, these methods usually assume that mechanism parameters such as the radius of the base circle have been previously determined, thereby producing a cam profile whose performance is as yet uncertain to a great extent. If, upon a thorough evaluation of the design at hand, this performance turns out to be unacceptable, then the relevant design parameters are changed using rules devised during years of design experience. Within the scope of this book, the design is evaluated through kinematic analysis, as discussed in Chapter 7. As a result of this analysis, the behavior of variables relevant to the mechanism operation can be studied. Variables that need to be monitored for flat-face followers are the contact-point eccentricity and the cam-profile curvature. The process of systematically choosing the mechanism parameters in order to produce a given displacement program of the follower, while minimizing the cam size and observing the pertinent constraints on mechanism variables, constitutes what is called *cam design optimization*. This subject has been given some attention (Chicurel, 1963; Mischke, 1970; Berzak, 1982; Buchsbaum and Freudenstein, 1983; Guoxun et al., 1988). Optimization methods based on spline functions were introduced in Angeles and López-Cajún (1983, 1984a and b, 1988), and were meant for flat-face and roller-followers of the translating and oscillating types while considering constraints on the pressure angle and contact-point eccentricity. Curvature constraints for flat-face followers were included in (Angeles et al., 1989a and b). Here and in Chapter 5, we expand upon the methods introduced in the aforementioned references and illustrate them with simple examples.

While cam-size minimization can be achieved using a straightforward procedure for the case of translating followers, oscillating followers are elusive to the same treatment. In fact, for the latter type of followers, a *direct* approach is introduced that aims at the minimization of the area of the cam profile by properly choosing the eccentricity of the follower face and the minimum value of the follower angle of rotation. This approach is direct in that it is based on the *first-order necessary conditions* that the mechanism parameters must satisfy in order to produce a stationary value of the cam area. However, the stationary values thus obtained do not always yield minima, and further investigation is necessary, as outlined in Section 4.3.

Furthermore, by virtue of the geometry of the follower of the mechanisms under study, the cam profiles for both translating and oscillating followers are *convex* everywhere. But this convexity brings about challenging related problems. Indeed, it is possible to choose values of the mechanism parameters yielding a cam profile that, while producing *theoretically* the prescribed motion of the follower (it satisfies the underlying kinematic constraints), is either structurally inadmissible or physically unfeasible. In fact, the profile synthesis procedure can yield curves that do not close, thereby being an unfeasible design, or curves that, although closed, contain cusps. Cusps are inadmissible because of the large contact stresses that they produce (Burr, 1981), and hence should be eliminated early in the design stage. In the case of translating followers, the conditions for cusp elimination are so simple as to allow for a straightforward determination of the base radius that, while producing a minimum-size cam profile, does so without cusps. However, for oscillating followers, the conditions for cusp elimination turn out to be somewhat more involved, thereby making impossible a straightforward determination of the parameter values leading to a minimum-size cam profile. Nevertheless, design guidelines are given that lead to a systematic determination of these parameters *interactively* with the aid of plotting software and a graphics monitor supplied with a *mouse*.

The flat-face translating and oscillating followers are discussed in this order, solid models of these mechanisms being displayed in Figs. 4.1.1 and 4.1.2. Numerical examples showing the applicability of the algorithms described are also included.

4.2 Translating Flat-Face Followers

4.2.1 Kinematic Relations

Let us consider Fig. 4.2.1, where OF and OC are lines attached to the fixed frame and to the cam, respectively. Moreover, ψ and $s(\psi)$ denote the angular displacement of the cam and the translational displacement of the follower. The latter can be considered as the sum of a constant, which in this case is the radius of the base circle, b, as yet to be determined, plus a *positive-semidefinite* function $\sigma(\psi)$, whose minimum value is zero, its maximum being equal to the total follower lift, h. A positive-semidefinite function, it is recalled, is a function that attains only positive or zero values within a given interval and cannot be zero everywhere in that interval. In our case, the interval of interest is $0 \leq \psi \leq 2\pi$. Thus,

$$s(\psi) = b + \sigma(\psi) \tag{4.2.1a}$$

and hence,

$$s'(\psi) = \sigma'(\psi), \qquad s''(\psi) = \sigma''(\psi) \tag{4.2.1b}$$

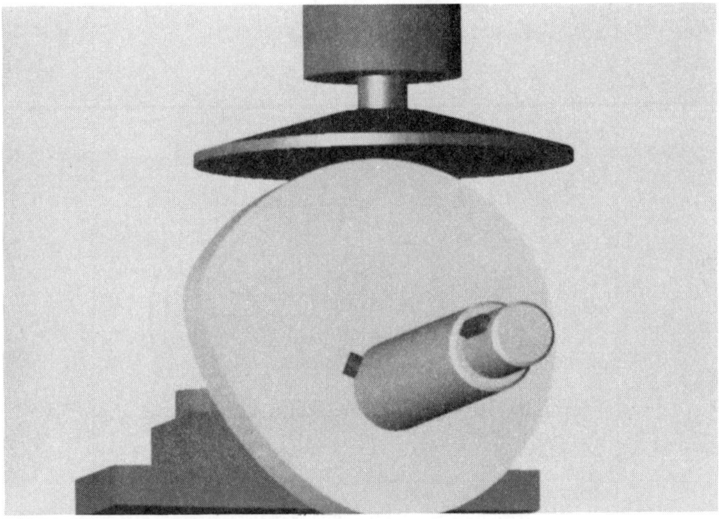

Fig. 4.1.1. Solid model of a planar cam mechanism with a translating follower (produced by Max A. González-Palacios).

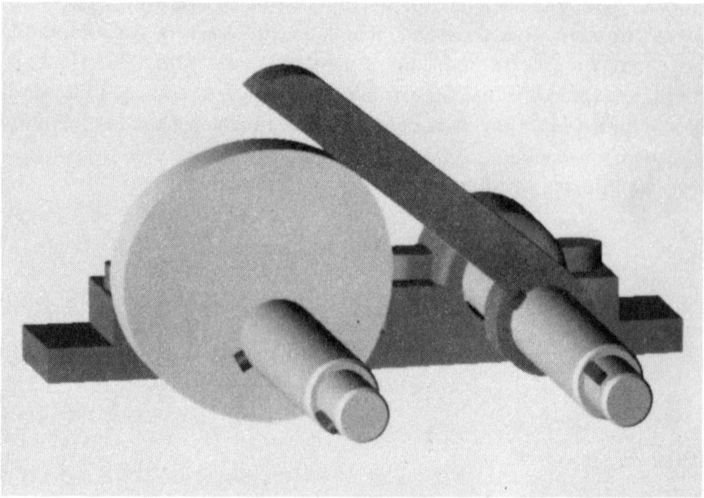

Fig. 4.1.2. Solid model of a planar cam mechanism with an oscillating follower (produced by Max A. González-Palacios).

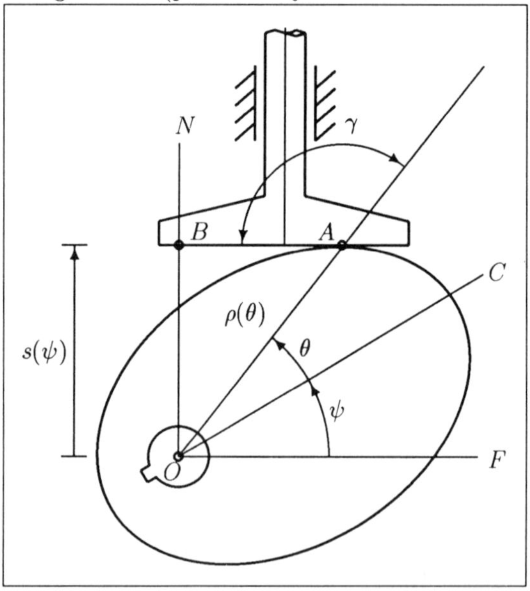

Fig. 4.2.1. Layout of a translating flat-face follower cam mechanism.

Furthermore, let θ and $\rho(\theta)$ be the polar coordinates defining the cam profile, and segment $\overline{BA}$, defined in Fig. 1.4.3 as ϵ, the eccentricity of the contact point A. Thus, ϵ is the lever arm of the normal component of the contact force between cam and follower, about the center of rotation of the cam, O. From the geometry of Fig. 4.2.1, one can write

$$s(\psi) = \rho(\theta)\sin(\psi+\theta) \tag{4.2.2}$$

Moreover, let v_C be the component of the velocity of the contact point A considered as a point of the cam disk, along the direction of translation of the follower, and v_F the velocity of the same point considered as a point of the follower. Since it is assumed that the cam and the follower remain in contact throughout the whole motion cycle, one can write

$$v_F = v_C \tag{4.2.3a}$$

where

$$v_F = \dot{s}(\psi) \tag{4.2.3b}$$

and

$$v_C = \rho(\theta)\,\dot{\psi}\,\cos(\psi+\theta) \tag{4.2.3c}$$

Hence,

$$\dot{s}(\psi) = \rho(\theta)\,\dot{\psi}\,\cos(\psi+\theta) \tag{4.2.3d}$$

or, if both sides are divided by the nonvanishing angular velocity of the cam, $\dot{\psi}$,

$$s'(\psi) = \rho(\theta)\cos(\psi+\theta) \tag{4.2.3e}$$

where the *chain rule* has been invoked, namely,

$$\dot{s}(\psi) \equiv \frac{ds}{dt} = \frac{ds}{d\psi}\frac{d\psi}{dt} = s'(\psi)\,\dot{\psi}$$

Combining eqs. (4.2.2) and (4.2.3*e*), expressions for ρ and θ as functions of ψ are readily derived, namely,

$$\rho(\psi) = \sqrt{s^2(\psi) + s'^2(\psi)} \tag{4.2.4}$$

and

$$\tan(\psi+\theta) = \frac{s(\psi)}{s'(\psi)} \tag{4.2.5a}$$

or

$$\theta(\psi) = \tan^{-1}\left[\frac{s(\psi)}{s'(\psi)}\right] - \psi \tag{4.2.5b}$$

Now, upon elimination of parameter ψ from eqs. (4.2.4) and (4.2.5), the parametric representation of the cam profile $\rho = \rho(\theta)$ is readily derived. Note that eq. (4.2.4) has the geometric interpretation shown in Fig. 4.2.2. Moreover, from the same figure and Fig. 4.2.1, it is apparent that the eccentricity of the contact point is identical to $s'(\psi)$. In fact, from Fig. 4.2.1,

$$\epsilon \equiv \overline{BA} = \rho(\theta)\cos(\psi + \theta) \tag{4.2.6}$$

Substitution of eq. (4.2.6) into eq. (4.2.3*d*), yields

$$\epsilon = \frac{\dot{s}}{\dot{\psi}} = \frac{ds}{d\psi} = s'(\psi) \tag{4.2.7}$$

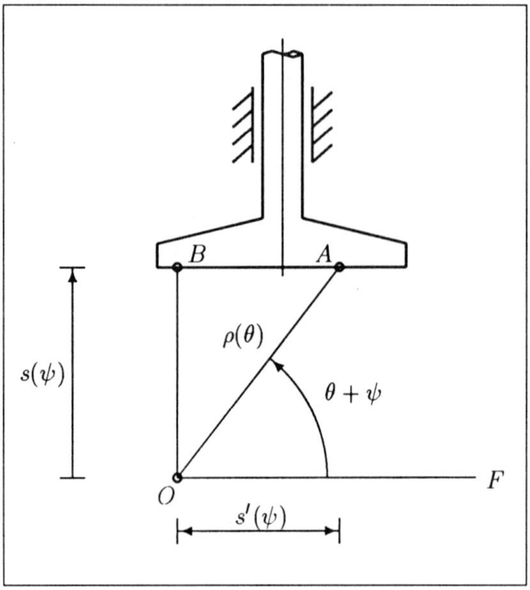

Fig. 4.2.2. Geometric interpretation of eq. (4.2.4).

Moreover, the Cartesian coordinates of a point of the cam profile, as functions of the cam angle of rotation ψ, are derived below. From basic polar-coordinate relations, we have

$$x = \rho\cos\theta, \qquad y = \rho\sin\theta$$

Upon expansion of eqs. (4.2.2) and (4.2.3*e*), one obtains, respectively,

$$\begin{aligned} s &= \rho\cos\theta\sin\psi + \rho\sin\theta\cos\psi \\ &= x\sin\psi + y\cos\psi \end{aligned} \tag{4.2.8a}$$

and

$$s' = \rho \cos\theta \cos\psi - \rho \sin\theta \sin\psi$$

$$= x \cos\psi - y \sin\psi \tag{4.2.8b}$$

Next, eqs. (4.2.8*a* & *b*) are rewritten in vector form as

$$\begin{bmatrix} \cos\psi & -\sin\psi \\ \sin\psi & \cos\psi \end{bmatrix} \begin{bmatrix} x \\ y \end{bmatrix} = \begin{bmatrix} s' \\ s \end{bmatrix} \tag{4.2.9a}$$

or

$$\mathbf{Q}(\psi)\,\mathbf{p}(\psi) = \mathbf{r}(\psi) \tag{4.2.9b}$$

where

$$\mathbf{Q}(\psi) = \begin{bmatrix} \cos\psi & -\sin\psi \\ \sin\psi & \cos\psi \end{bmatrix}, \quad \mathbf{p}(\psi) = \begin{bmatrix} x(\psi) \\ y(\psi) \end{bmatrix}, \quad \mathbf{r}(\psi) = \begin{bmatrix} s'(\psi) \\ s(\psi) \end{bmatrix} \tag{4.2.9c}$$

matrix $\mathbf{Q}(\psi)$ being orthogonal, i.e.,

$$\mathbf{Q}^T\mathbf{Q} = \mathbf{Q}\,\mathbf{Q}^T = \mathbf{1}, \quad \det(\mathbf{Q}) = +1 \tag{4.2.9d}$$

Furthermore, $\mathbf{1}$ denotes the 2×2 identity matrix. Hence, eq. (4.2.9*b*) can readily be solved for vector $\mathbf{p}(\psi)$, namely,

$$\mathbf{p}(\psi) = \mathbf{Q}^T(\psi)\,\mathbf{r}(\psi) = \begin{bmatrix} s(\psi)\sin\psi + s'(\psi)\cos\psi \\ s(\psi)\cos\psi - s'(\psi)\sin\psi \end{bmatrix} \tag{4.2.10a}$$

or

$$x = s(\psi)\sin\psi + s'(\psi)\cos\psi \tag{4.2.10b}$$

$$y = s(\psi)\cos\psi - s'(\psi)\sin\psi \tag{4.2.10c}$$

Note that $\mathbf{r}(\psi)$ represents the coordinates of a point of the cam profile in a coordinate system fixed to the machine frame, whereas $\mathbf{p}(\psi)$ represents those of the same point in a coordinate system fixed to the cam disk.

The determination of the radius of the optimum base circle under constraints on the eccentricity ϵ, is pursued in the section below.

4.2.2 Cam-Profile Optimization Under Eccentricity Constraints

Our aim in this chapter is to produce cam profiles for the flat-face followers that are of minimum size. We would like first to derive those minimum-area contours disregarding any constraints, introducing these later in the design, as needed. It turns out, however, that, for translating flat-face followers, unconstrained minimum-size contours yield a vanishing base radius, which is impractical. In fact, below we show that the area of the cam profile grows monotonically with the base radius b and hence, the minimizing value of b becomes zero, which is not realizable. In fact, the area of the cam disk, as discussed in Subsection 3.3.1, is given by

$$A = \frac{1}{2} \int_0^{2\pi} \rho^2(\psi)\, \theta'(\psi)\, d\psi \tag{4.2.11}$$

where $\rho(\psi)$ is given by eq. (4.2.4), and $\theta'(\psi)$ is obtained below. Combining eqs. (4.2.1a) and (4.2.3e), one has

$$\tan(\theta + \psi) = s(\psi)/s'(\psi) \tag{4.2.12}$$

Differentiation of both sides of eq. (4.2.12) with respect to ψ yields

$$[1 + \theta'(\psi)] \sec^2(\theta + \psi) = \frac{1}{s'(\psi)} \left[s'(\psi) - \frac{s(\psi)}{s'(\psi)} s''(\psi) \right] \tag{4.2.13}$$

but $\sec^2(\theta + \psi) = 1 + \tan^2(\theta + \psi)$. Therefore, if eq. (4.2.5a) is recalled,

$$\sec^2(\theta + \psi) = 1 + \frac{s^2(\psi)}{s'^2(\psi)} \tag{4.2.14}$$

Upon substituting eq. (4.2.14) into eq. (4.2.13) and reducing the ensuing expressions, $\theta'(\psi)$ is obtained as

$$\theta'(\psi) = -\frac{s(\psi) + s''(\psi)}{s^2(\psi) + s'^2(\psi)} \tag{4.2.15}$$

Substitution of eqs. (4.2.15) and (4.2.4) into eq. (4.2.11) yields

$$A = \frac{1}{2} \int_0^{2\pi} s(\psi)\, [\, s(\psi) + s''(\psi)\,]\, d\psi \tag{4.2.16}$$

Furthermore, from eqs. (4.2.1a & b),

$$s = b + \sigma(\psi), \qquad s''(\psi) = \sigma''(\psi) \tag{4.2.17}$$

Substitution of eq. (4.2.17) into eq. (4.2.16) leads to

$$A = \pi b^2 + A_1 b + A_2 + A_3 \tag{4.2.18a}$$

where

$$A_1 \equiv \int_0^{2\pi} \sigma(\psi)\, d\psi$$

$$A_2 \equiv \frac{1}{2}\int_0^{2\pi} \sigma^2(\psi)\, d\psi \tag{4.2.18b}$$

$$A_3 \equiv \frac{1}{2}\int_0^{2\pi} \sigma(\psi)\, \sigma''(\psi)\, d\psi$$

Note that, by virtue of the positive-semidefiniteness of $\sigma(\psi)$, $A_1 > 0$. Moreover, A_2 is obviously positive definite as well, whereas A_3 turns out to be negative, as shown below. In fact,

$$A_3 \equiv \frac{1}{2}\int_0^{2\pi} \sigma(\psi)\, \sigma''(\psi)\, d\psi = \frac{1}{2}\int_{\sigma'_a}^{\sigma'_b} \sigma(\psi)\, d\sigma'(\psi) \tag{4.2.19}$$

where σ'_a and σ'_b are the integration limits under the adopted variable of integration, $\sigma'(\psi)$. Next, the last integral is evaluated by parts, namely,

$$A_3 = \sigma(\psi)\, \sigma'(\psi)\Big|_0^{2\pi} - \frac{1}{2}\int_{\sigma_a}^{\sigma_b} \sigma'(\psi)\, d\sigma(\psi) \tag{4.2.20}$$

σ_a and σ_b being the integration limits for the corresponding integration variable, $\sigma(\psi)$, while the first term of the right-hand side of the latter expression vanishes by virtue of the periodicity of both $\sigma(\psi)$ and $\sigma'(\psi)$. Hence,

$$A_3 = -\frac{1}{2}\int_{\sigma_a}^{\sigma_b} \sigma'(\psi)\, d\sigma(\psi) = -\frac{1}{2}\int_0^{2\pi} \sigma'^2(\psi)\, d\psi \tag{4.2.21}$$

which is obviously positive as well. Thus, the area A becomes

$$A = \pi b^2 + A_1 b + A_0 \tag{4.2.22}$$

where $A_0 \equiv A_2 + A_3$. Hence, A_0 is the sum of a positive and a negative quantity, its sign being indefinite, for it depends on the normal motion at hand, $\sigma(\psi)$. In summary, A is the sum of a quadratic expression in b with positive coefficients and a sign-indefinite term independent of b. Therefore, its minimum for $b \geq 0$ is attained at $b = 0$, the said minimum being A_0. The conclusion is that, in the absence of constraints, the minimum-size cam profile is obtained in this case for a vanishing base radius, a solution that is not very practical. However, when constraints are included, a nonzero optimum base radius is derived.

We proceed first to the cam-profile optimization under eccentricity constraints. A more elaborate optimization, based on a curvature constraint, is to be discussed in Subsection 4.2.3.

Since ϵ is the lever arm of the moment applied by the vertical component of the contact force onto the follower, its value should be bounded in order to prevent the occurrence of large bending moments. Thus, a *bound* ϵ_M is imposed on ϵ, i.e.,

$$\epsilon \leq \epsilon_M \tag{4.2.23}$$

From eq. (4.2.7) it is apparent that, for a prescribed follower-motion program, the eccentricity will also be fixed. However, what is important is not the maximum eccentricity itself but its value with respect to the radius of the base circle, b. Hence, we will specify that the latter, as yet to be determined, be a given multiple of ϵ_M, i.e.,

$$b = \nu\,\epsilon_M \tag{4.2.24}$$

and hence, ν is the ratio of the base-circle radius to the maximum allowed eccentricity ϵ_M. Moreover, the value ν appearing in eq. (4.2.24) is usually dictated or constrained by the design environment. Substitution of eqs. (4.2.7) and (4.2.24) into eq. (4.2.23) leads to

$$s'(\psi) \leq b/\nu \tag{4.2.25}$$

From relation (4.2.25), for a prescribed maximum eccentricity, the left-hand side attains a maximum value when the equality is satisfied. Therefore, the minimum radius of the base circle b_ϵ, is obtained as

$$b_\epsilon = M\,\nu \tag{4.2.26a}$$

where

$$M \equiv \max_{\psi}\{|\,s'(\psi)\,|\}, \qquad \psi_1 \leq \psi \leq \psi_2 \tag{4.2.26b}$$

and $\Delta\psi \equiv \psi_2 - \psi_1$ is the cam angle of rotation for the rise or the return phases, depending on which is under analysis. Thus, the kinematic problem of cam-profile optimization, for any displacement program, has been reduced to finding the point—i.e., the value of ψ in the rise or the return phase—where the follower velocity attains its maximum absolute value, that is, the point where the acceleration vanishes. Equation (4.2.26a) gives the optimum radius of the base circle, i.e., the minimum-size cam disk for a given follower-motion program that obeys the eccentricity constraint (4.2.23), for a prescribed ratio ν.

Note that both the rise and the return phases should be considered. Obviously, the phase with the largest absolute value of $s'(\psi)$ will dominate the design. In particular, if both phases are defined based on the same normal motion introduced in Section 2.2, then the phase with the shortest duration will contain the largest maximum. Moreover, in quick-return motions, the phase with the shortest duration is the return phase.

4.2.3 Example 4.2.1: Minimum-Size Cam Under a Maximum-Eccentricity Constraint

Determine the minimum base radius b_ϵ for a cam profile to produce a rise of $h = 50$ mm. The rise phase has a duration of $\Delta\psi = 2\pi/3$, and the return phase a duration of $\pi/3$. Use *i)* a cycloidal motion and *ii)* a 3-4-5 polynomial motion, and constrain the maximum eccentricity to be at most equal to b_ϵ.

Solution:

We have, in general,

$$s'(\psi) = \sigma'(\psi) = \frac{h}{\Delta\psi}\,\tau'(x)$$

from which it is apparent that we must design for the return phase, which has the smaller $\Delta\psi$ and hence, yields the larger $s'(\psi)$.

i) Cycloidal Motion

In this case,

$$\tau'(x) = 1 - \cos 2\pi x$$

and hence,

$$\max_x\{|\,\tau'(x)\,|\} = 2, \qquad \text{for} \quad x = 0.5$$

Now, we substitute the given numerical values, with $\Delta\psi = \pi/3$, thereby obtaining

$$M = \frac{50}{\pi/3} \times 2 = \frac{300}{\pi} = 95.5 \text{ mm}$$

In our case, $\nu = 1$ and hence,

$$b_\epsilon = M = 95.5 \text{ mm}$$

ii) 3-4-5 Polynomial Motion

Now we have, from eqs. (2.3.19)

$$\tau'(x) = 30\,x^2\,(x-1)^2$$

and hence, as found in Subsection 2.3.7,

$$\max_x\{|\,\tau'(x)\,|\} = 1.875$$

which thus yields, upon substitution of the given numerical values,

$$M = \frac{50}{\pi/3} \times 1.875 = 89.52 \text{ mm}$$

and, thus,

$$b_\epsilon = M = 89.52 \text{ mm}$$

i.e., the base radius b_ϵ is 91% bigger than the rise h if a cycloidal motion is used, but only 79% bigger than the rise if a 3-4-5 polynomial motion is used.

4.2.4 Curvature Constraint

In the previous section the optimum value of the radius of the base circle was found based on a simple constraint on the contact-point eccentricity. However, as we recall in this section, the curvature of the profile may become unbounded, and hence, the profile may contain cusps, in the presence of relatively short durations of the rise or the return phase. Conditions under which the curvature becomes unbounded are derived below:

The curvature κ of the cam profile is now obtained using eq. (3.2.16), reproduced below for quick reference

$$\kappa = \text{sgn}\,[\,\lambda'(q)\,]\,\frac{\mathbf{p}''^T(q)\,\mathbf{E}\,\mathbf{p}'(q)}{||\,\mathbf{p}'(q)\,||^3} \tag{4.2.27a}$$

In order to evaluate κ using eq. (4.2.27a), we must first decide how to choose q. From the discussion of the previous subsections, it is natural to choose $q = \psi$. In this case, however,

$$\lambda'(\psi) < 0$$

as we will show presently. We do this with a special case, namely, an *eccentric* cam, that is shown in Fig. 4.2.3. In that figure, the cam profile is a circular disk of radius a that rotates about a point O a distance e from its center. In this case,

$$\lambda = a\left(\frac{\pi}{2} - \psi\right)$$

and hence,

$$\lambda'(\psi) = -a < 0$$

Thus, κ will take on the general expression shown below

$$\kappa = -\frac{\mathbf{p}''^T(\psi)\,\mathbf{E}\,\mathbf{p}'(\psi)}{\|\,\mathbf{p}'(\psi)\,\|^3} = \frac{\mathbf{p}'^T(\psi)\,\mathbf{E}\,\mathbf{p}''(\psi)}{\|\,\mathbf{p}'(\psi)\,\|^3} \tag{4.2.27b}$$

where $\mathbf{p}'(\psi)$ and $\mathbf{p}''(\psi)$ are derived from differentiation of $\mathbf{p}(\psi)$ as given by eq. (4.2.10a), i.e.,

$$\mathbf{p}'(\psi) = \mathbf{Q}'^T(\psi)\,\mathbf{r}(\psi) + \mathbf{Q}^T(\psi)\,\mathbf{r}'(\psi) \tag{4.2.28a}$$

or

$$\mathbf{p}'(\psi) = \mathbf{Q}^T(\psi)\,\mathbf{Q}(\psi)\,\mathbf{Q}'^T(\psi)\,\mathbf{r}(\psi) + \mathbf{Q}^T(\psi)\,\mathbf{r}'(\psi) \tag{4.2.28b}$$

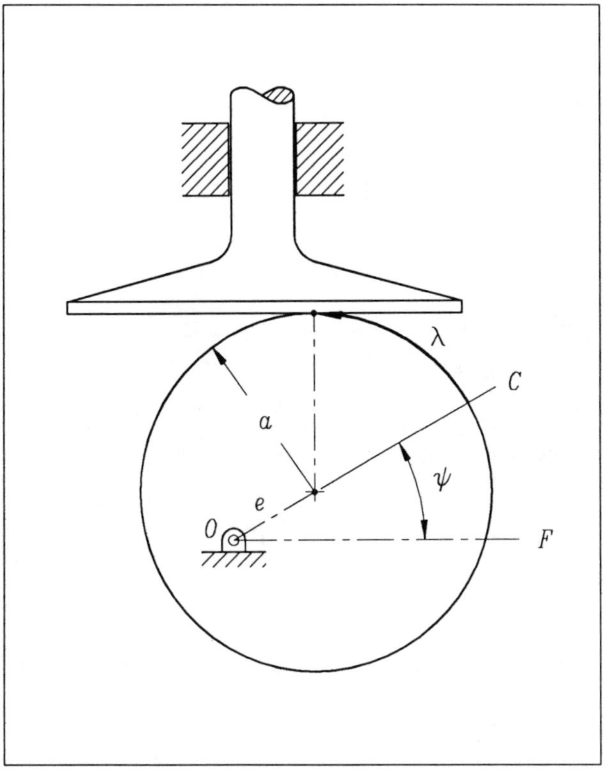

Fig. 4.2.3. An eccentric cam.

the first term of the right-hand side of eq. (4.2.28a) appears to be multiplied by the identity matrix in the form of $\mathbf{Q}^T(\psi)\,\mathbf{Q}(\psi)$. However,

$$\mathbf{Q}(\psi)\,\mathbf{Q}'^T(\psi) = \begin{bmatrix} 0 & 1 \\ -1 & 0 \end{bmatrix} \equiv -\mathbf{E} \tag{4.2.28c}$$

$\mathbf{E}$ being the orthogonal matrix introduced in Section 3.2.2. Hence, eq. (4.2.28b) can be rewritten as

$$\mathbf{p}'(\psi) = \mathbf{Q}^T(\psi)\,[\,\mathbf{r}'(\psi) - \mathbf{E}\,\mathbf{r}(\psi)\,] \tag{4.2.28d}$$

Further differentiation of $\mathbf{p}'(\psi)$ with respect to ψ yields

$$\mathbf{p}''(\psi) = \mathbf{Q}'^T(\psi)\,[\,\mathbf{r}'(\psi) - \mathbf{E}\,\mathbf{r}(\psi)\,] + \mathbf{Q}^T(\psi)\,[\,\mathbf{r}''(\psi) - \mathbf{E}\,\mathbf{r}'(\psi)\,]$$

or, similar to eq. (4.2.28b),

$$\begin{aligned}\mathbf{p}''(\psi) &= \mathbf{Q}^T(\psi)\,\mathbf{Q}(\psi)\,\mathbf{Q}'^T(\psi)\,[\,\mathbf{r}'(\psi) - \mathbf{E}\,\mathbf{r}(\psi)\,] + \mathbf{Q}^T(\psi)\,[\,\mathbf{r}''(\psi) - \mathbf{E}\,\mathbf{r}'(\psi)\,] \\ &= \mathbf{Q}^T(\psi)\,[\,\mathbf{r}''(\psi) - \mathbf{E}\,\mathbf{r}'(\psi) - \mathbf{E}\,\mathbf{r}'(\psi) + \mathbf{E}^2\mathbf{r}(\psi)\,]\end{aligned}$$

But,

$$\mathbf{E}^2 = -\mathbf{1}$$

and hence,

$$\mathbf{p}''(\psi) = \mathbf{Q}^T(\psi)\,\left[\,\mathbf{r}''(\psi) - 2\,\mathbf{E}\,\mathbf{r}'(\psi) - \mathbf{r}(\psi)\,\right] \tag{4.2.29}$$

Moreover, $\mathbf{r}'(\psi)$ and $\mathbf{r}''(\psi)$, required in eqs. (4.2.28d) and (4.2.29), are readily derived by differentiation of $\mathbf{r}(\psi)$ as given by eqs. (4.2.9c), i.e.,

$$\mathbf{r}' = \begin{bmatrix} s'' \\ s' \end{bmatrix}, \qquad \mathbf{r}'' = \begin{bmatrix} s''' \\ s'' \end{bmatrix} \tag{4.2.30}$$

the argument ψ having been dropped for brevity, and all derivatives being understood to be with respect to ψ. Applying the formulas for curvature as given by eq. (4.2.27), one obtains

$$\kappa = \frac{1}{s + s''} \tag{4.2.31a}$$

and hence, the radius of curvature r becomes

$$r = s + s'' \tag{4.2.31b}$$

From eq. (4.2.31a), it is clear that the curvature will be infinite, i.e., the cam profile will have a cusp, whenever the radius of curvature vanishes, that is, when

$$s + s'' = 0 \tag{4.2.31c}$$

in full agreement with the result derived by Cowie (1961) using geometric arguments. For example, the displacement $s(\psi)$ produced by the eccentric cam of Fig. 4.2.3,

$$s(\psi) = a + e\sin\psi, \quad s'(\psi) = e\cos\psi, \quad s''(\psi) = -e\sin\psi$$

and hence,

$$r = s(\psi) + s''(\psi) = a$$

Now, recalling eq. (4.2.1a), eq. (4.2.31c) can be rewritten as

$$\sigma''(\psi) + \sigma(\psi) + b = 0 \tag{4.2.32}$$

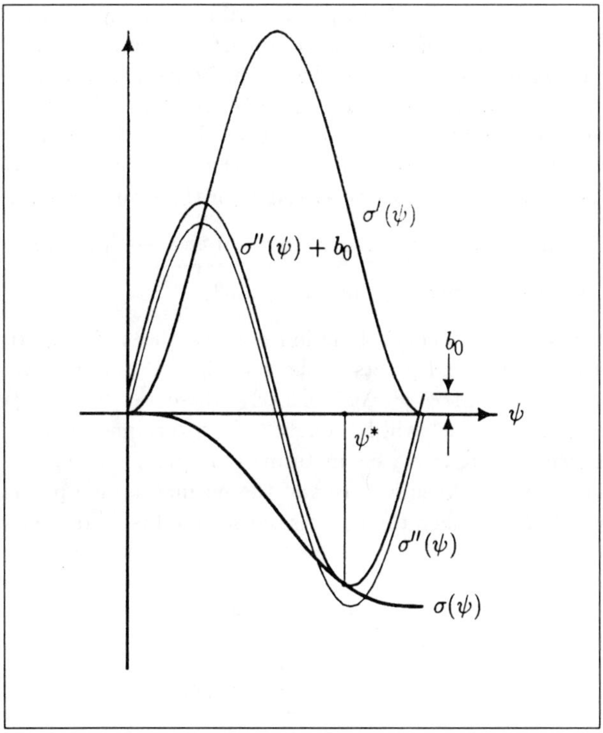

Fig. 4.2.4. Geometric interpretation of eq. (4.2.32).

Since we are interested in the minimum value of b, denoted by b_0, we will impose the additional condition that, at a position defined by ψ^*, the curve $\sigma''(\psi)$ becomes tangent to $-[\,b+\sigma(\psi)\,]$, as shown in Fig. 4.2.4.

Obviously, eq. (4.2.32) also holds at $\psi = \psi^*$, with $b = b_0$, i.e.,

$$\sigma''(\psi^*) + \sigma(\psi^*) + b_0 = 0 \tag{4.2.33}$$

At this point, we have two unknowns to determine, namely, b_0 and ψ^* in order to verify eq. (4.2.31c) or eq. (4.2.32). Hence, a second equation is needed to determine these unknowns. This equation is obtained by noticing that the negative of the slope of $\sigma(\psi)$, i.e., $-\sigma'(\psi)$, should be equal to the slope of $\sigma''(\psi)$ at the point where tangency occurs, namely, at $\psi = \psi^*$, which yields

$$\sigma'''(\psi^*) = -\sigma'(\psi^*) \tag{4.2.34}$$

We now have two equations, eqs. (4.2.33) and (4.2.34), to determine the two unknowns b_0 and ψ^*. Thus, the minimum value of b, b_0, under curvature constraints, can be found interactively with the aid of a graphics terminal supplied with a *mouse*—for a description of this, see Chapter 9. To do this, function $\sigma(\psi)$ is shifted up and down by an amount b until a point is found at which the curve $\sigma(\psi) + b$ is tangent to the curve $-\sigma''(\psi)$. The value of ψ at the point of tangency is the value sought, ψ^*, and the minimum value of b, b_0, is that producing tangency.

Alternatively, the problem can be solved algorithmically, namely,

1. find ψ^* as the point where the curves $\sigma'(\psi)$ and $-\sigma'''(\psi)$ intersect.
2. with ψ^* known, compute b_0 from eq. (4.2.33).

Note that, if a smooth-enough function $\sigma(\psi)$ has been chosen, then both $\sigma'(\psi)$ and $\sigma'''(\psi)$ vanish at the end points of the rise phase or, alternatively, the descent phase, i.e., at $\psi = 0$ and $\psi = \Delta\psi$. Clearly, these values of ψ are trivial, and hence, intermediate values ψ^* will be sought. Moreover, function $\sigma(\psi)$ is apparently problem-dependent, for it is stated in terms of a particular follower-motion program. If this motion is defined ir terms of the normal motion program, as defined in eqs. (2.2.1) and (2.2.2), then the expressions for the first three derivatives of $\sigma(\psi)$ become

$$\sigma'(\psi) = \frac{h}{\Delta\psi}\,\tau'(x) \tag{4.2.35a}$$

$$\sigma''(\psi) = \frac{h}{(\Delta\psi)^2}\,\tau''(x) \tag{4.2.35b}$$

$$\sigma'''(\psi) = \frac{h}{(\Delta\psi)^3}\,\tau'''(x) \tag{4.2.35c}$$

which are more suitable expressions, for they are given in terms of normal-motion derivatives. These are readily available from a catalog of functions, namely, those discussed in Section 2.3. Additionally, only two problem-dependent parameters, h and $\Delta\psi$, appear in the above derivatives. Thus, eqs. (4.2.32) and (4.2.34) take on the forms

$$\tau''(x^*) + (\Delta\psi)^2 [\tau(x^*) + b_0/h] = 0 \qquad (4.2.36a)$$

and

$$\tau'''(x^*) + (\Delta\psi)^2 \tau'(x^*) = 0 \qquad (4.2.36b)$$

Note that eq. (4.2.36*b*), which determines the value x^* that provides ψ^* as $x^*\Delta\psi$, is independent of h. Thus, ψ^* does not depend on the rise value h.

4.2.5 Example 4.2.2: Cam Profile for Cycloidal Motion

Determine the cam profile of a cam mechanism with a translating flat-face follower so as to avoid cusps. The total rise is 50 mm, and the displacement program is given in Table 4.2.1. Consider that the rise and the return phases are synthesized using the normal cycloidal motion. This is given as

$$\tau = x - \frac{1}{2\pi}\sin 2\pi x, \quad 0 \le x \le 1 \qquad (4.2.37)$$

Table 4.2.1. Displacement program for Example 4.2.2.

Phase	Cam angle of rotation	Displacement (mm)
D_1 (dwell)	22.50°	0
R_1 (rise)	157.50°	+50
D_2 (dwell)	60.00°	0
R_2 (return)	120.00°	−50

Solution:

First, we will determine the optimum base radius for cusp avoidance based on the methodology outlined above. To this end, we compute the three derivatives needed, namely,

$$\tau' = 1 - \cos 2\pi x \qquad (4.2.38a)$$

$$\tau'' = 2\pi \sin 2\pi x \qquad (4.2.38b)$$

$$\tau''' = 4\pi^2 \cos 2\pi x \qquad (4.2.38c)$$

Substitution of eqs. (4.2.38a & c) into eq. (4.2.36b) yields,

$$4\pi^2 \cos 2\pi x + (\Delta\psi)^2(1 - \cos 2\pi x) = 0 \tag{4.2.39}$$

or

$$\cos 2\pi x = -\frac{f^2}{1-f^2}, \qquad f \equiv \frac{\Delta\psi}{2\pi} \tag{4.2.40}$$

From eq. (4.2.40), it is clear that the right-hand side will always be negative since $0 < f < 1$. Therefore, two values of x, x_1^* and x_2^*, are to be obtained, which satisfy eq. (4.2.40) as illustrated in Fig. 4.2.5. From that figure, these values of x^* lie in the intervals given below:

$$x_1^* \in [\,0.25,\ 0.50\,], \qquad x_2^* \in [\,0.50,\ 0.75\,]$$

Furthermore,

$$x_1^* + x_2^* = 1$$

Thus, two values of b_0 are derived, namely,

$$b_0 = \begin{cases} -h\,[\,F(f) + x^*\,], & \text{for } x^* \in [\,0.25,\ 0.50\,] \\ -h\,[\,-F(f) + x^*\,], & \text{for } x^* \in [\,0.50,\ 0.75\,] \end{cases} \tag{4.2.41a}$$

where

$$F(f) \equiv \frac{\sqrt{1-2f^2}}{2\pi f^2} \tag{4.2.41b}$$

Clearly, $F(f)$ is defined in the interval $0 < f \le \sqrt{2}/2$, since $F(0) \to \infty$ and the radical of eq. (4.2.41b) becomes negative for $f > \sqrt{2}/2$.

Now, since $x*$ and b_0/h are positive definite, the second value of b_0/h, as given by eq. (4.2.41a), produces

$$x^* = F(f) - b_0/h \tag{4.2.42}$$

On the other hand, from eq. (4.2.40), for $x = x^*$,

$$x^* = \frac{1}{2\pi}\cos^{-1}\left(-\frac{f^2}{1-f^2}\right) \tag{4.2.43}$$

where $\cos^{-1}(\cdot)$ is understood to be the smaller of the two possible values in the interval $[\,0,\ 2\pi\,]$.

Combination of eqs. (4.2.42) and (4.2.43) leads to

$$\frac{b_0}{h} = \frac{1}{2\pi}\left[2\pi F(f) - \cos^{-1}\left(-\frac{f^2}{1-f^2}\right)\right] \tag{4.2.44}$$

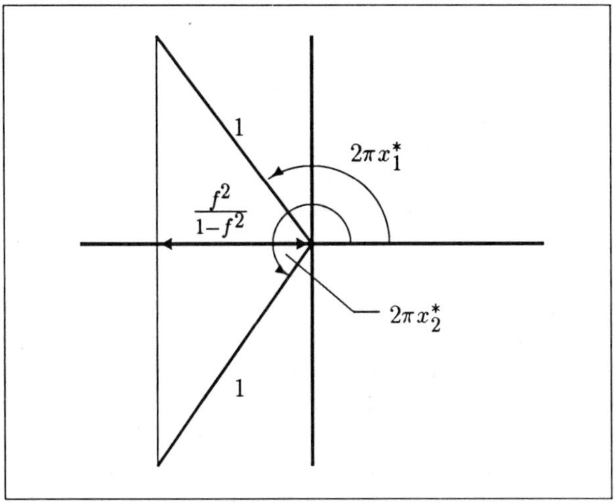

Fig. 4.2.5. Geometric representation of eq. (4.2.40).

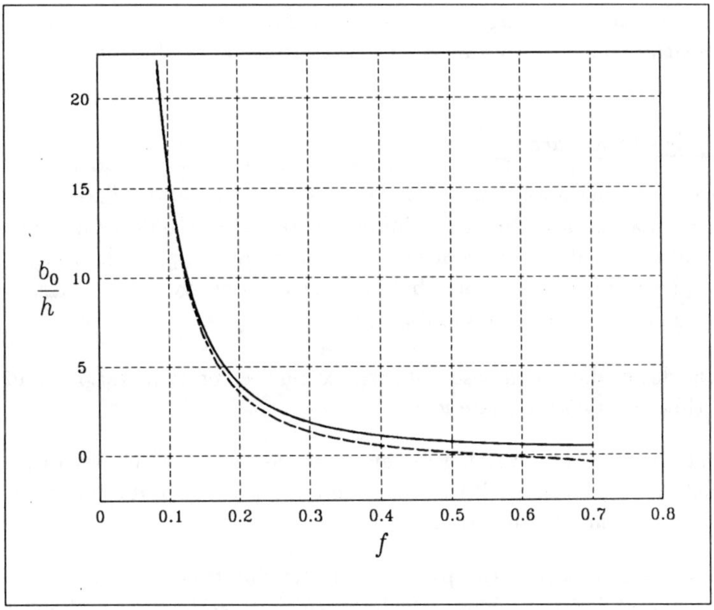

Fig. 4.2.6. Plot of eq. (4.2.44) for cycloidal motion.

Clearly, eq. (4.2.44) produces two branches for b_0/h, one for each of the two values of the $\cos^{-1}(\cdot)$ function. These are displayed in Fig. 4.2.6. The solid line corresponds to values of x_2^*, whereas the dashed line corresponds to values of x_1^*. The resulting values of b_0/h for $f < 0.085$ are omitted because of their impracticality. Of course, this plot is only valid if the displacement program is given as a cycloidal motion. Furthermore, from the solid line displayed in Fig. 4.2.6, the designer can select, for a given ratio f, the corresponding ratio b_0/h. Otherwise, this ratio is obtained directly as explained below.

Now, for the case at hand, the return takes place in $\Delta\psi = 2\pi/3$, and $f = 1/3$. This value yields $\cos 2\pi x^* = -1/8$, and hence, $x_1^* = 0.2700$ and $x_2^* = 0.7300$. Correspondingly, $b_0/h = -1.533$ and $b_0/h = 0.5333$ and therefore, the optimum value is $b_0 = 26.66$ mm. Once the base radius is determined, the Cartesian coordinates are computed using eqs. (4.2.10*b* & *c*). The cam profile obtained is shown in Fig. 4.2.7.

If the return phase took place in $\pi/3$ rad, i.e., half the angle of the given value of Example 4.2.2, then, $f = 1/6$ and $\cos 2\pi x = -1/35$, from which x_1^* and x_2^* are computed as $x_1^* = 0.2546$ and $x_2^* = 0.7454$, the minimum radius b_0 being found from eq. (4.2.44) as

$$b_0/h = 4.823$$

That is, we obtain a very large cam, with a base radius of almost five times the rise. In particular, for the given value of h (= 50 mm), $b_0 = 241.15$ mm.

4.2.6 Design Procedure

The above example shows that, if both the rise and the return phases are based on the same normal motion, then the minimum base radius should be determined from the phase of shorter duration. Since the cam in the return phase is usually under no load, this phase is the shorter one, but this need not be always the case.

The design procedure is summarized below:

1. If the design dictates a base-radius/maximum-eccentricity ratio ν, find b_ϵ based on this; otherwise, assume $\nu = 1$.

2. Next, verify if the foregoing design is free of cusps, namely, by displaying the function $s''(\psi) + s(\psi)$. If this function does not change its sign in the interval of interest, then the design is cusp-free

3. Otherwise, determine the phase of shorter duration, whether the rise or the return, and find the minimum value b_0/h that can be used without producing cusps.

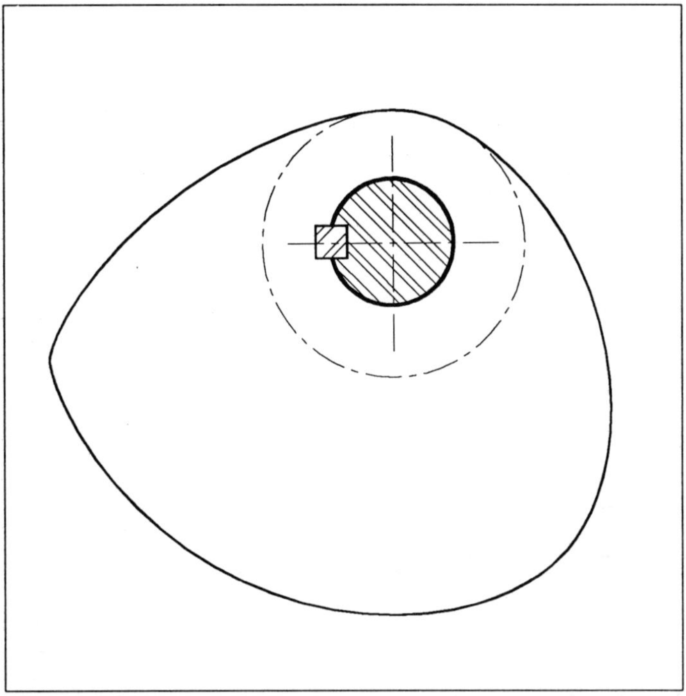

Fig. 4.2.7. Cam profile for the translating flat-face follower.

4.3 Oscillating Flat-Face Followers

4.3.1 Kinematic Relations

It will be assumed that the follower-displacement program is given as the sum of a constant β, a parameter to be determined, and a positive-semidefinite function $\sigma(\psi)$, whose minimum value is 0, its maximum value being $\Delta\phi$, the amplitude of the follower oscillation, i.e.,

$$\phi(\psi) = \beta + \sigma(\psi), \quad 0 \leq \psi \leq 2\pi \tag{4.3.1}$$

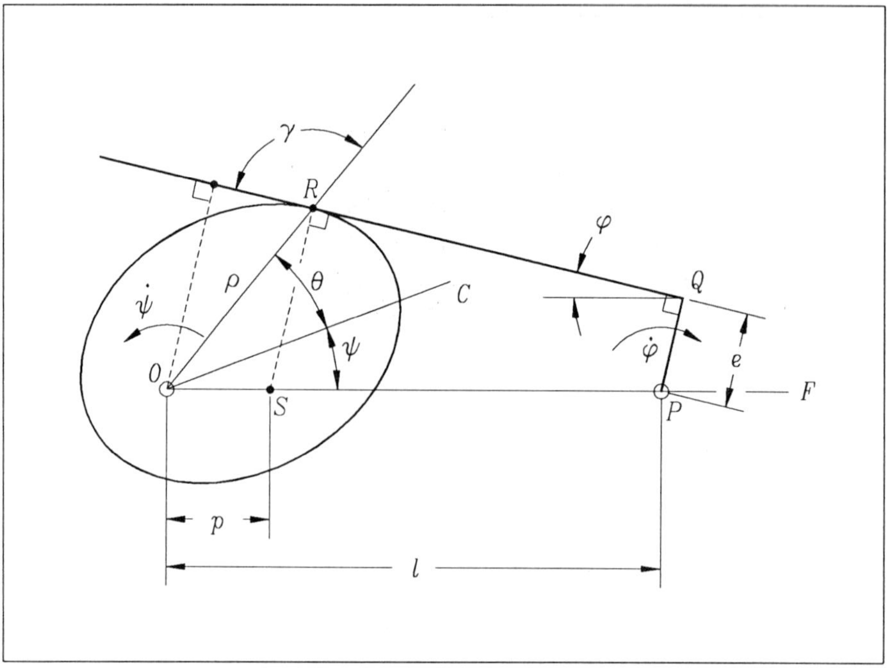

Fig. 4.3.1. Layout of a disk cam with oscillating flat-face follower.

The layout of a cam mechanism with an oscillating flat-face follower is shown in Fig. 4.3.1. Let OF and OC be lines attached to the fixed frame and to the cam plate, respectively. Moreover, e is the follower offset—segment $\overline{PQ}$—, a parameter to be determined, whereas θ and $\rho(\theta)$—segment $\overline{OR}$—are the polar coordinates describing the cam profile, and ψ and ϕ are the angular displacements of the cam plate and the follower, respectively. Constant β of function $\phi(\psi)$, as well as e, are to be determined by aiming at a cusp-free, minimum-area cam profile. This procedure is outlined below.

From triangle OSR of Fig. 4.3.1, and upon application of the *sines law*, we obtain

$$\frac{u}{\sin\left(\gamma - \frac{\pi}{2}\right)} = \frac{\rho}{\sin\left(\frac{\pi}{2}\right)} \tag{4.3.2}$$

But, from the same figure,

$$\gamma = \pi - (\psi + \phi + \theta) \tag{4.3.3a}$$

and hence,

$$\sin\left(\gamma - \frac{\pi}{2}\right) \equiv \cos\left(\psi + \phi + \theta\right) \tag{4.3.3b}$$

Upon substitution of eq. (4.3.3b) into eq. (4.3.2), we obtain

$$\rho \cos(\psi + \phi + \theta) = u \cos\phi \tag{4.3.4}$$

What we need now is an expression for u in terms of the mechanism parameters and the motion variables, $\phi(\psi)$ and its time derivatives. This relation is readily derived by application of the Aronhold-Kennedy Theorem (Hunt, 1978). Indeed, since points O and P are, correspondingly, the *instant centers* of the cam and the follower with respect to the mechanism frame, the instant center of the follower with respect to the cam is located along line OF. Moreover, this instant center lies along the common normal N to the two profiles. The intersection of line OF with the normal N, point S in Fig. 4.3.1, is the instant center of the follower with respect to the cam—or the other way around. As a consequence, the velocity of point S, regarded as a point of the cam, $\dot{\mathbf{s}}_C$, equals that of the same point when considered a point of the follower. The velocity of the latter is denoted by $\dot{\mathbf{s}}_F$. Moreover, both $\dot{\mathbf{s}}_C$ and $\dot{\mathbf{s}}_F$ are perpendicular to line OP, and, for positive values of $\dot{\psi}$ and $\dot{\phi}$, as indicated in the same figure, point upwards. Upon equating these two velocities, we derive the desired relation, namely,

$$u\dot{\psi} = (\ell - u)\dot{\phi}$$

from which we can readily solve for u, i.e.,

$$u = \ell \frac{\phi'(\psi)}{1 + \phi'(\psi)} \tag{4.3.5a}$$

and hence, u can attain both positive and negative values, depending on $\phi'(\psi)$, with negative values of u meaning that point S lies to the left of point O. Furthermore, we have used the relation

$$\phi'(\psi) \equiv \frac{d\phi}{d\psi} = \frac{\dot{\phi}}{\dot{\psi}} \tag{4.3.5b}$$

Now, eq. (4.3.4) can be rewritten as

$$\rho \cos(\psi + \phi + \theta) = \ell \frac{\phi' \cos\phi}{1 + \phi'} \tag{4.3.6}$$

where the argument ψ has been omitted from $\phi'(\psi)$, for brevity.

Likewise, from the geometry of the quadrilateral $OPQR$ and, upon equating the projection of segment $\overline{OR}$ with the sum of the projections of segments $\overline{OP}$ and $\overline{PQ}$ onto the normal N of Fig. 4.3.1, we can readily derive the relation

$$\rho \sin(\psi + \phi + \theta) = \ell \sin\phi + e \tag{4.3.7}$$

Now, eqs. (4.3.6) and (4.3.7) are the *synthesis equations* that allow us to determine the cam profile for given values of parameters e and ℓ, and a given displacement program, $\phi(\psi)$. Indeed, the polar coordinates $(\rho,\ \theta)$ of the cam profile are readily found by suitably combining eqs. (4.3.6) and (4.3.7). Thus, from the sum of the squares of both sides of those equations, we obtain

$$\rho = \frac{\sqrt{\ell^2\,\phi'^2\cos^2\phi + [\,(1+\phi')\,(\ell\,\sin\phi + e)\,]^2}}{1+\phi'} \tag{4.3.8a}$$

Furthermore, upon division of each side of eq. (4.3.7) by the corresponding side of eq. (4.3.6), we derive

$$\theta = \tan^{-1}\left[\frac{(1+\phi')\,(\ell\,\sin\phi + e)}{\ell\,\phi'\cos\phi}\right] - (\psi + \phi) \tag{4.3.8b}$$

Alternatively, we can define the Cartesian coordinates $(x,\ y)$ of the cam profile as shown below. First, we define

$$\mu \equiv \psi + \phi \tag{4.3.9}$$

and next, we rewrite eqs. (4.3.6) and (4.3.7) as

$$\rho\,\cos\theta\,\cos\mu - \rho\,\sin\theta\,\sin\mu = \ell\,\frac{\phi'\cos\phi}{1+\phi'} \tag{4.3.10a}$$

$$\rho\,\cos\theta\,\sin\mu + \rho\,\sin\theta\,\cos\mu = \ell\,\sin\phi + e \tag{4.3.10b}$$

from which we readily identify the Cartesian coordinates sought, namely,

$$x = \rho\,\cos\theta, \qquad y = \rho\,\sin\theta \tag{4.3.10c}$$

Now, eqs. (4.3.10*a* & *b*) can be rewritten in vector form, for ease of manipulation, as

$$\mathbf{Q}(\mu)\,\mathbf{p}(\psi) = \mathbf{r}(\psi) \tag{4.3.11a}$$

with the definitions given below:

$$\mathbf{Q}(\mu) = \begin{bmatrix} \cos\mu & -\sin\mu \\ \sin\mu & \cos\mu \end{bmatrix} \tag{4.3.11b}$$

$$\mathbf{p}(\psi) = \begin{bmatrix} x(\psi) \\ y(\psi) \end{bmatrix}, \qquad \mathbf{r}(\psi) = \begin{bmatrix} \xi(\psi) \\ \eta(\psi) \end{bmatrix} \tag{4.3.11c}$$

where $\xi(\psi)$ and $\eta(\psi)$ have the geometric interpretation given in Fig. 4.3.2 and correspond to the right-hand sides of eqs. (4.3.10a & b), namely,

$$\xi \equiv \ell \frac{\phi' \cos\phi}{1+\phi'}, \qquad \eta \equiv \ell \sin\phi + e \tag{4.3.11d}$$

Moreover, ξ and η are the coordinates of the contact point R with respect to a frame centered at O and maintaining a fixed orientation with respect to the follower.

Since matrix $\mathbf{Q}(\mu)$ is orthogonal, solving for $\mathbf{p}(\psi)$ from eq. (4.3.11a) reduces to transposing $\mathbf{Q}(\mu)$, namely,

$$\mathbf{p}(\psi) = \mathbf{Q}^T(\mu)\,\mathbf{r}(\psi) \tag{4.3.12a}$$

i.e.,

$$x(\psi) = \ell \frac{\phi' \cos\phi}{1+\phi'} \cos\mu + (\ell \sin\phi + e) \sin\mu \tag{4.3.12b}$$

$$y(\psi) = -\ell \frac{\phi' \cos\phi}{1+\phi'} \sin\mu + (\ell \sin\phi + e) \cos\mu \tag{4.3.12c}$$

where μ can be readily calculated from eq. (4.3.9), once $\phi(\psi)$ is known.

While eqs. (4.3.12b & c) allow the calculation of the Cartesian coordinates of actual points of the cam profile, it will prove convenient to introduce non-dimensional variables, as shown below:

$$\bar{e} \equiv \frac{e}{\ell}, \quad \bar{x}(\psi) \equiv \frac{x(\psi)}{\ell}, \quad \bar{y}(\psi) \equiv \frac{y(\psi)}{\ell}, \quad \bar{\xi}(\psi) \equiv \frac{\xi(\psi)}{\ell}, \quad \bar{\eta}(\psi) \equiv \frac{\eta(\psi)}{\ell} \tag{4.3.13}$$

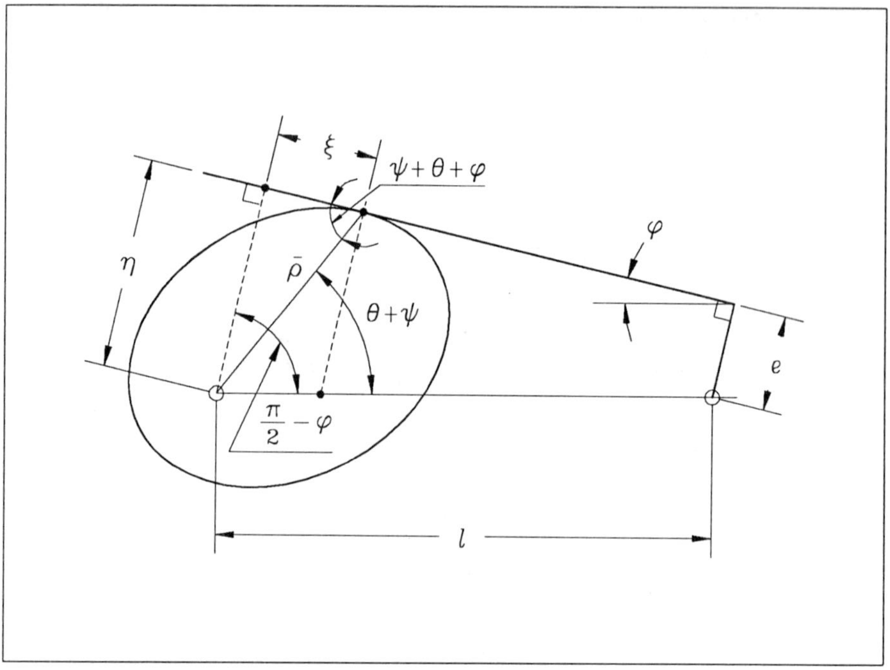

Fig. 4.3.2. Geometric interpretation of eqs. (4.3.11*d*).

and hence, eqs. (4.3.12*b* & *c*) lead to expressions for the non-dimensional coordinates of the cam profile, namely,

$$\bar{x}(\psi) = \frac{\phi' \cos\phi}{1+\phi'} \cos(\psi+\phi) + (\sin\phi + \bar{e}) \sin(\psi+\phi) \tag{4.3.14a}$$

$$\bar{y}(\psi) = -\frac{\phi' \cos\phi}{1+\phi'} \sin(\psi+\phi) + (\sin\phi + \bar{e}) \cos(\psi+\phi) \tag{4.3.14b}$$

Vectors $\mathbf{p}(\psi)$ and $\mathbf{r}(\psi)$, introduced in eqs.(4.3.11*c*), are now redefined as their non-dimensional counterparts, namely,

$$\mathbf{p}(\psi) \equiv \begin{bmatrix} \bar{x}(\psi) \\ \bar{y}(\psi) \end{bmatrix}, \qquad \mathbf{r} \equiv \begin{bmatrix} \bar{\xi}(\psi) \\ \bar{\eta}(\psi) \end{bmatrix} \tag{4.3.14c}$$

The *shape* of the profile is totally determined by $\bar{e}$ and the motion program, the actual size of the cam being determined by the *scaling* parameter ℓ. However, from eq. (4.3.1), $\phi(\psi)$ contains parameter β, which must also be found. In summary, then, we have two parameters, $\bar{e}$ and β, to minimize the size of the *non-dimensional cam*. This minimization is pursued, first, under no constraints, in Subsection 4.3.2, with curvature constraints being imposed in Subsection 4.3.3.

4.3.2 Unconstrained Cam-Size Minimization

In order to obtain the optimum unconstrained parameters, β and $\bar{e}$, we attempt the minimization of the area of the cam disk. Similar to the case of the translating follower, the area is given by

$$A = \frac{1}{2}\int_{\theta_0}^{\theta_0+2\pi} \bar{\rho}^2\, d\theta = \frac{1}{2}\int_0^{2\pi} \bar{\rho}^2(\psi)\,\theta'(\psi)\, d\psi \tag{4.3.15}$$

Contrary to the case of the translating follower, however, in this case the area is no longer a monotonic function of the design parameters β and $\bar{e}$. Hence, the area admits non-trivial stationary points, which are found as described below. Differentiation of both sides of eq. (4.3.8*b*) with respect to ψ yields

$$\theta'(\psi) = \frac{\phi'^2(1+\phi')(1+\bar{e}\sin\phi) - \phi''\cos\phi\,(\sin\phi+\bar{e})}{\phi'^2\cos^2\phi + (\sin\phi+\bar{e})^2(1+\phi')^2} - (1+\phi') \tag{4.3.16}$$

where the trigonometric identity

$$\cos^2\phi = \frac{1}{1+\tan^2\phi}$$

has been used. Substitution of eqs. (4.3.8*a*) and (4.3.16) into eq. (4.3.15) leads to

$$A = Z_1 + Z_2 + Z_3 \tag{4.3.17}$$

where

$$Z_1 = \int_0^{2\pi} \frac{\phi'^2}{1+\phi'} \sin(\beta+\sigma)\,[\sin(\beta+\sigma)+\bar{e}\,]\, d\psi$$

$$Z_2 = -\int_0^{2\pi} \frac{\phi''}{(1+\phi')^2} \cos(\beta+\sigma)\,[\sin(\beta+\sigma)+\bar{e}\,]\, d\psi$$

$$Z_3 = -\int_0^{2\pi} [\sin(\beta+\sigma)+\bar{e}\,]^2(1+\phi')\, d\psi$$

The foregoing equations show an explicit relationship between the area A and the parameters β and $\bar{e}$. In order to find the values of these parameters that minimize A, the necessary conditions are introduced, namely,

$$\frac{\partial A}{\partial \beta} = 0, \qquad \frac{\partial A}{\partial \bar{e}} = 0 \tag{4.3.18}$$

After algebraic manipulations, eqs. (4.3.18) lead to

$$U_3 \sin 2\beta + U_4 \cos 2\beta + \bar{e}\,(-U_1 \sin\beta + U_2 \cos\beta) = 0 \tag{4.3.19a}$$

$$2\pi\bar{e} + U_2 \sin\beta + U_1 \cos\beta = 0 \tag{4.3.19b}$$

where

$$U_1 = \int_0^{2\pi} A_1\, s_1\, d\psi - \int_0^{2\pi} A_2\, c_1\, d\psi - 2\int_0^{2\pi} A_3\, s_1\, d\psi \tag{4.3.20a}$$

$$U_2 = \int_0^{2\pi} A_1\, c_1\, d\psi + \int_0^{2\pi} A_2\, s_1\, d\psi - 2\int_0^{2\pi} A_3\, c_1\, d\psi \tag{4.3.20b}$$

$$U_3 = \int_0^{2\pi} A_1\, c_2\, d\psi - \int_0^{2\pi} A_2\, s_2\, d\psi - \int_0^{2\pi} A_3\, c_2\, d\psi \tag{4.3.20c}$$

$$U_4 = \int_0^{2\pi} A_1\, s_2\, d\psi - \int_0^{2\pi} A_2\, c_2\, d\psi - \int_0^{2\pi} A_3\, s_2\, d\psi \tag{4.3.20d}$$

and

$$s_1 \equiv \sin\sigma(\psi), \quad c_1 \equiv \cos\sigma(\psi), \quad s_2 \equiv \sin 2\sigma(\psi), \quad c_2 \equiv \cos 2\sigma(\psi)$$

with coefficients $A_i(\psi)$, for $i = 1,\ 2,\ 3$, being defined as

$$A_1(\psi) = \frac{\phi'^2}{1+\phi'}, \quad A_2(\psi) = \frac{\phi''}{(1+\phi')^2}, \quad A_3(\psi) = 1 + \phi' \tag{4.3.20e}$$

From eq. (4.3.19b),

$$\bar{e} = -\frac{U_2 \sin\beta + U_1 \cos\beta}{2\pi} \tag{4.3.21}$$

Upon substitution of eq. (4.3.21) into eq. (4.3.19a), one obtains

$$B_1 \cos^2\beta + B_2 \sin^2\beta + B_3 \sin\beta\cos\beta = 0 \tag{4.3.22}$$

with coefficients B_i for $i = 1,\ 2,\ 3$, being defined as

$$B_1 = 2\pi U_4 - U_1 U_2, \quad B_2 = -B_1, \quad B_3 = 4\pi U_3 - U_2^2 + U_1^2$$

Now, we resort to the trigonometric relationships

$$\sin\beta = \frac{2t}{1+t^2}, \quad \cos\beta = \frac{1-t^2}{1+t^2}$$

where

$$t \equiv \tan\frac{\beta}{2}$$

and hence, eq. (4.3.22) reduces to

$$t^4 - C\,t^3 - 6\,t^2 + C\,t + 1 = 0 \tag{4.3.23a}$$

where C is computed as

$$C = \frac{2B_3}{B_1} \tag{4.3.23b}$$

Thus, the problem has been reduced to finding the roots of a quartic equation, eq. (4.3.23a), which can be done explicitly, using Ferrari's method (Archbold, 1970; Lelong-Ferrand and Arnaudiès, 1978). Ferrari's method is based on a suitable change of variable leading to a cubic equation, which in turn can be solved using Tartaglia and Cardan's method. Alternatively, roots of eq. (4.3.23a) can be obtained either by resorting to a root-finding routine applicable to polynomials, or graphically by plotting the polynomial and pointing its intersections with the t axis using a *mouse*—see Chapter 9. Once β has been found, the eccentricity $\bar{e}$ can be computed *uniquely* using either eq.(4.3.19a) or (4.3.19b), thereby completing the design. In deciding which of the two foregoing equations to use, one should look for the most *robust* for the purpose at hand, i.e., the one in which $\bar{e}$ appears with the coefficient bearing the larger absolute value. Hence, a safe choice would be to use the second of those equations all the time.

However, some problems arise when following this procedure: *i*) Equation (4.3.23a) admits both real and complex roots; *ii*) the real roots may yield not only local minima, but also local maxima or saddle points; and *iii*) even if a minimum is obtained, this may yield a profile with a cusp or, even worse, a curve that does not close. Below we elaborate on how to determine if a profile has a cusp and discuss means of producing cusp-free profiles.

4.3.3 Curvature Constraints

Here, we derive the condition under which cusps occur. Note that cusps can be detected by simple inspection of the cam profile, but sometimes the presence of a cusp may be uncertain. Hence, an algebraic criterion allowing us to detect unambiguously the presence of cusps, which will be also used to prevent them, is warranted. The formula for the curvature, as given by eq. (3.2.17)—where $\mathrm{sgn}[\lambda'(q)] = -1$ as explained in Subsection 4.2.4—, is recalled and written here for quick reference, namely,

$$\kappa = \frac{\mathbf{p}'^T(\psi)\mathbf{E}\,\mathbf{p}''(\psi)}{\|\mathbf{p}'(\psi)\|^3} \tag{4.3.24}$$

Now, $\mathbf{p}'$ and $\mathbf{p}''$ are derived. Differentiation of both sides of eq. (4.3.11a) with respect to ψ yields

$$\mathbf{p}'(\psi) = \mu'(\psi)\,\mathbf{Q}'^T(\mu)\,\mathbf{r}(\psi) + \mathbf{Q}^T(\mu)\,\mathbf{r}'(\psi) \tag{4.3.25a}$$

where, from the definition of μ,

$$\mu'(\psi) = 1 + \phi'(\psi) = 1 + \sigma'(\psi) \tag{4.3.25b}$$

Premultiplying the first term of the right-hand side by the identity matrix given in the form $\mathbf{Q}^T\mathbf{Q}$ and reducing the ensuing expressions, we obtain

$$\mathbf{p}'(\psi) = \mathbf{Q}^T(\mu)\,[\mathbf{r}'(\psi) - \mu'(\psi)\,\mathbf{E}\,\mathbf{r}(\psi)] \tag{4.3.26}$$

where $\mathbf{E}$ is the matrix already defined in Subsection 3.2.2, namely,

$$\mathbf{E} = \begin{bmatrix} 0 & -1 \\ 1 & 0 \end{bmatrix}$$

Upon differentiation of both sides of eq. (4.3.26) with respect to ψ, we obtain

$$\mathbf{p}''(\psi) = \mathbf{Q}^T(\mu)\,[\mathbf{r}''(\psi) - \mu''(\psi)\,\mathbf{E}\,\mathbf{r}(\psi) - 2\,\mu'(\psi)\,\mathbf{E}\,\mathbf{r}'(\psi) - \mu'^2(\psi)\,\mathbf{r}(\psi)] \tag{4.3.27}$$

where $\mu''(\psi) = \phi''(\psi) = \sigma''(\psi)$.

From eq. (4.3.14c), $\mathbf{r}$ and its derivatives are

$$\mathbf{r} = \begin{bmatrix} \bar{\xi} \\ \bar{\eta} \end{bmatrix}, \quad \mathbf{r}'(\psi) = \begin{bmatrix} \bar{\xi}'(\psi) \\ \bar{\eta}'(\psi) \end{bmatrix}, \quad \mathbf{r}''(\psi) = \begin{bmatrix} \bar{\xi}''(\psi) \\ \bar{\eta}''(\psi) \end{bmatrix} \tag{4.3.28a}$$

Now,

$$\mathbf{r}'(\psi) - \mu'(\psi)\,\mathbf{E}\,\mathbf{r} = \begin{bmatrix} \bar{\xi}'(\psi) + \mu'(\psi)\,\bar{\eta} \\ \bar{\eta}'(\psi) - \mu'(\psi)\bar{\xi} \end{bmatrix}$$

But

$$\bar{\eta}'(\psi) - \mu'(\psi)\bar{\xi} = 0 \tag{4.3.28b}$$

and hence, the foregoing expression reduces to

$$\mathbf{r}'(\psi) - \mu'(\psi)\,\mathbf{E}\,\mathbf{r} = \begin{bmatrix} \bar{\xi}'(\psi) + \mu'(\psi)\,\bar{\eta} \\ 0 \end{bmatrix}$$

Thus,

$$\mathbf{p}'(\psi) = \mathbf{Q}^T(\mu) \begin{bmatrix} \bar{\xi}'(\psi) + \mu'(\psi)\,\bar{\eta} \\ 0 \end{bmatrix} \tag{4.3.29a}$$

and

$$\mathbf{p}''(\psi) = \mathbf{Q}^T(\mu) \begin{bmatrix} \bar{\xi}''(\psi) + \mu''(\psi)\bar{\eta} + \mu'(\psi)\,\bar{\eta}'(\psi) \\ -\mu'(\psi)\,[\,\bar{\xi}'(\psi) + \mu'(\psi)\bar{\eta}\,] \end{bmatrix} \tag{4.3.29b}$$

Substitution of eqs. (4.3.29*a* & *b*) into eq. (4.3.24), after reducing the ensuing relations, yields

$$\kappa = \frac{\mu'(\psi)}{|\bar{\xi}'(\psi) + \mu'(\psi)\bar{\eta}|} \tag{4.3.30a}$$

From the definition of μ and $\bar{\xi}$, i.e.,

$$\mu = \psi + \phi, \qquad \bar{\xi} = \frac{\phi' \cos\phi}{1 + \phi'} \tag{4.3.30b}$$

their derivatives are readily computed, namely,

$$\mu'(\psi) = 1 + \phi'(\psi) \tag{4.3.30c}$$

$$\bar{\xi}'(\psi) = -\frac{\phi'^2(1 + \phi')\sin\phi + \phi''\cos\phi}{[1 + \phi']^2} \tag{4.3.30d}$$

Upon substitution of the above equations into eq. (4.3.30) and after simplifications, one obtains

$$\kappa = \frac{(1 + \phi')^3}{|\,F(\psi)\,|} \tag{4.3.31a}$$

where

$$F(\psi) \equiv \phi''\cos\phi + (1 + \phi')\,[\,(1 + 2\phi')\sin\phi + \bar{e}\,(1 + \phi')^2\,] \tag{4.3.31b}$$

From eq. (4.3.31*b*), it is apparent that κ becomes negative if $1+\phi'(\psi) < 0$. However, changes in curvature sign are not physically possible for this type of cams and hence, we must constrain κ to be positive, i.e., $\phi'(\psi)$ must obey

$$1+\phi'(\psi) > 0 \tag{4.3.31c}$$

Moreover, κ becomes unbounded if $F(\psi)$, defined in eq. (4.3.31*b*), vanishes. Geometrically, this means that the cam profile has a cusp, which is undesirable because of the large contact stresses that it produces. Thus, we will have to enforce the condition that $F(\psi)$ does not have a change of sign. Since we do not know beforehand if $F(\psi)$ is either positive- or negative-definite, we express the condition for this function to be sign-definite by saying that its global maximum and its global minimum have the same sign, i.e.,

$$\min_{\psi}\{F(\psi)\}\max_{\psi}\{F(\psi)\} > 0 \tag{4.3.31d}$$

which follows by virtue of the continuity of $F(\psi)$.

A bound on $\bar{e}$ guaranteeing that $F(\psi)$ does not vanish was found, in terms only of $\phi'(\psi)$ and $\phi''(\psi)$, in (Angeles et al. 1989b). This bound, however, is too conservative and can lead to very large cams. A less conservative approach is recommended below. It is necessary to find suitable values of β and $\bar{e}$ that render function $F(\psi)$ sign-definite. The search for suitable values of β and $\bar{e}$ can be implemented *interactively* with a monitor display of function $F(\psi)$ and a menu-driven change of parameters β and $\bar{e}$ with the aid of a *mouse*, as outlined below.

Based on condition (4.3.31c), the radius of curvature r—that is, the reciprocal of the curvature κ—can be expressed as

$$r = \left|\frac{F(\psi)}{(1+\phi')^3}\right| \tag{4.3.32a}$$

the argument of the absolute-value function in the latter equation now being expanded as

$$\frac{F(\psi)}{(1+\phi')^3} = g(\psi) + \bar{e} \tag{4.3.32b}$$

with $g(\psi)$ defined as

$$g(\psi) \equiv \frac{\phi''\cos\phi}{(1+\phi')^3} + \frac{(1+2\phi')\sin\phi}{(1+\phi')^2} \tag{4.3.32c}$$

where we have separated the two design parameters, $\bar{e}$ and β, for $g(\psi)$ contains only β, in ϕ, but is independent of $\bar{e}$. The condition for cusp avoidance can now be formulated as the nonvanishing of the right-hand side of eq.(4.3.32*b*). This condition allows us to determine $\bar{e}$ for a given value of β, as explained below.

Now, we let $g_m(\beta)$ and $g_M(\beta)$ denote the *global* minimum and the *global* maximum of $g(\psi;\beta)$, for a given value of β, and choose $\bar{e}$ so that

$$[g_m(\beta)+\bar{e}][g_M(\beta)+\bar{e}]>0 \tag{4.3.33}$$

Furthermore, the nondimensional base radius $\bar{b}$ for this cam is given by

$$\bar{b}=\sin\beta+\bar{e} \tag{4.3.34}$$

which must be positive. Moreover, let $\bar{c}$ denote the non-dimensional value of $\rho(\theta)$ at the upper dwell, i.e.,

$$\bar{c}=\sin(\beta+\Delta\phi)+\bar{e} \tag{4.3.35}$$

which also must be positive. Note that whether $\bar{b}$ and $\bar{c}$ is larger depends on both β and $\Delta\phi$. That is, in this case, the base radius is not necessarily the minimum value of $\rho(\theta)$.

On the other hand, from the form of $g(\psi)$, as given by eq.(4.3.32c),

$$g(\psi;\beta+\pi)=-g(\psi;\beta) \tag{4.3.36}$$

changing β to $\beta+\pi$ *and* reversing the sign of $\bar{e}$ do not affect the radius of curvature r given in eq.(4.3.324a). By the same token, the same changes in β and $\bar{e}$ affect neither the polar coordinates of the cam profile, given by eqs.(4.3.8a & b), nor its Cartesian coordinates, eqs.(4.3.12b & c). As a consequence, we only need to search for β in the interval $0\leq\beta\leq\pi$. However, $\bar{e}$ can be either positive or negative but, in any event, we will constrain it to lie between -1 and $+1$, for absolute values of $\bar{e}$ greater than unity are bound to yield disproportionately large follower arms.

In order to minimize the size of the cam, then, we will try to render $\bar{b}$ and $\bar{c}$ as low as possible. The application of the foregoing concepts to the design of the cam profile is first summarized and then illustrated with an example below.

4.3.4 Design Procedure

The design procedure can now be summarized in algorithmic form, namely,

1) The duration of the phases D_1, R_1, D_2, R_2, the maximum angular displacement $\Delta\phi$, and the motion type $\sigma(\psi)$ are input.

2) Based either on a catalog of normal motions or on a spline-generated motion, the displacement, velocity, and acceleration curves are produced.

3) Stationary values of $\bar{e}$ and β are computed from the real roots of eq. (4.3.23a).

4) With $\bar{e}$ and β known, function $F(\psi)$, as given by eq. (4.3.31b), is now evaluated. If this function does not change sign and the cam size is acceptable, i.e., the values of $\bar{b}$, $\bar{c}$ and $\bar{e}$ are all smaller than unity, then the profile has no cusps and is feasible. Hence, the design is complete.

5) Otherwise, suitable values of β and $\bar{e}$ are found that render $F(\psi)$ sign-definite. A menu-driven mode and a mouse are highly recommended here.

6) Once $\bar{e}$ and β are available, $\theta(\psi)$ and $\bar{\rho}(\psi)$ or, alternatively, the Cartesian coordinates $\bar{x}(\psi)$ and $\bar{y}(\psi)$ are evaluated.

7) ℓ is chosen based on space availability and stress considerations. Once ℓ is known, the actual mechanism is fully designed.

4.3.5 Example 4.3.1: Cam-Profile Determination

Synthesize a cam follower mechanism for an oscillating flat-face follower, which will produce the follower displacement program appearing in Table 4.3.1. Assume that the rise and the return phases are cycloidal.

Table 4.3.1: Prescribed angular-displacement program.

Phase	Angle of rotation(ψ)	Displacement ($\Delta\phi$)
D_1(dwell)	36°	0°
R_1(rise)	144°	+15°
D_2(dwell)	54°	0°
R_2(return)	126°	−15°

Solution:

First, an unconstrained minimization is attempted. This requires computing coefficient C of the quartic equation (4.3.23*a*), which is given in eq.(4.3.23*b*). To calculate this value, we need coefficients B_1 and B_3 that are given in turn in terms of coefficients U_k, for $k = 1, \ldots, 4$. These coefficients are calculated using expressions (4.3.20*a*–*d*), that require the calculation of six integrals. These were calculated using Romberg's method, which produced the values shown below:

$$U_1 = -1.69244, \quad U_2 = -10.9704, \quad U_3 = -5.15283, \quad U_4 = -1.63323$$

The computed value of C turned out to be

$$C = -12.6428$$

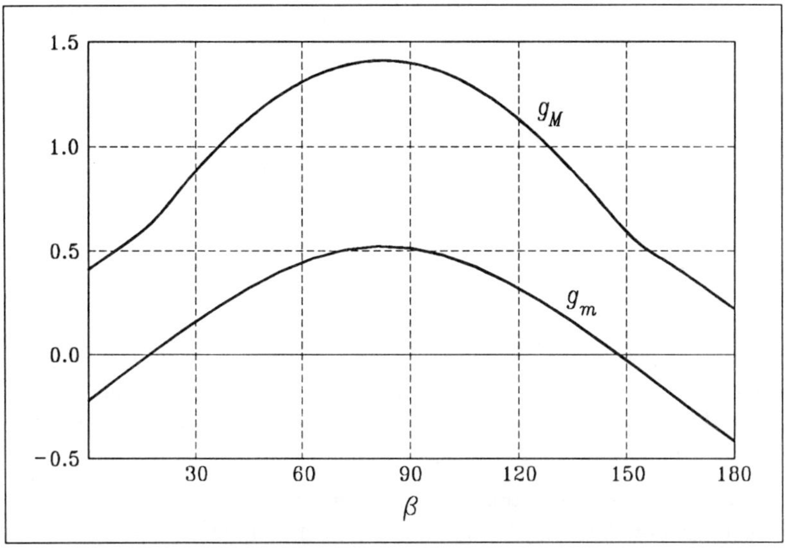

Fig. 4.3.3. Minimum and maximum values of $g(\psi; \beta)$ vs. β.

Next, the roots of the quartic equation (4.3.23a) were obtained with the aid of *MATHEMATICA* (Wolfram, 1988), which produced:

$$t_1 = -1.1663, \qquad t_2 = 0.07675, \quad t_3 = 0.8574, \quad t_4 = 13.03$$

From these roots, four pairs of design parameters β and $\bar{e}$ were computed, namely,

$$\begin{aligned} \beta_1 &= -98.47°, & \bar{e}_1 &= -1.7666 \\ \beta_2 &= -8.78°, & \bar{e}_2 &= -0.00026 \\ \beta_3 &= 81.22°, & \bar{e}_3 &= 1.7665 \\ \beta_4 &= 171.2°, & \bar{e}_4 &= 0.0003124 \end{aligned}$$

In order to work with values of β in the interval $[0, \pi]$, we add π, or 180°, to the first two values of β given above, and change the sign of their associated $\bar{e}$-values, namely,

$$\begin{aligned} \beta_1 &= 81.53°, & \bar{e} &= 1.7666 \\ \beta_2 &= 171.2°, & \bar{e} &= 0.00026 \end{aligned}$$

Furthermore, the values of $\bar{b}$ and $\bar{c}$ were calculated for each of the above pairs, namely,

$$\begin{aligned} \bar{b}_1 &= 2.7557, & \bar{c}_1 &= 2.7595 \\ \bar{b}_2 &= 0.1532, & \bar{c}_2 &= -0.1077 \\ \bar{b}_3 &= 2.7557, & \bar{c}_3 &= 2.7595 \\ \bar{b}_4 &= 0.1532, & \bar{c}_4 &= -0.1077 \end{aligned}$$

which amount to basically only two different designs. Of these, the first one is discarded, for it leads to profiles that are too large—both $\bar{b}$ and $\bar{c}$ are greater than 1—, while the second one is unfeasible because it yields a negative value of $\rho(\theta)$ at the upper dwell.

Thus, the unconstrained minimization approach is not applicable here. We will now find a feasible design yielding a small cam profile using interactive graphics. First, we plot $g_m(\beta)$ and $g_M(\beta)$, which are displayed in Fig. 4.3.3. From this figure, it is apparent that no cusps occur in the interval $-17° \leq \beta \leq 145°$, in which both $g_m(\beta)$ and $g_M(\beta)$ are positive, if we let $\bar{e} = 0$. In fact, even negative values of $\bar{e}$ are possible here. For example, if we choose $\beta = 90°$, we can let $\bar{e} = -0.5$ and still satisfy the no-cusp condition, eq.(4.3.33). In this case, the numerical values of $\bar{b}$ and $\bar{c}$ would be $\bar{b} = 0.5$ and $\bar{c} = 0.4659$. While these values are quite acceptable, they were obtained at the expense of a large absolute value of $\bar{e}$, namely, 50% the distance between the centers of rotation of the cam and the follower. If we want a lower value of $|\bar{e}|$, we can use the same plot, Fig. 4.3.3, to search for a more suitable design. Note that, from its definition, a lower value of $\bar{b}$ requires not only a small absolute value of $\bar{e}$, but also small values of $\sin\beta$ and $\sin(\beta + \Delta\phi)$. The above-mentioned plot then leads us to the extreme of the previously defined interval, $17° \leq \beta \leq 145°$, namely, $\beta = 17°$ and $\beta = 145°$. These yield the values of $\bar{b}$ and $\bar{c}$ displayed below:

$$\begin{aligned} &\bar{e} = 0, \quad \beta = 17°, \quad \bar{b} = 0.2924, \quad \bar{c} = 0.5299 \\ &\bar{e} = 0, \quad \beta = 145°, \quad \bar{b} = 0.5736, \quad \bar{c} = 0.3420 \end{aligned}$$

Clearly, a smaller cam is obtained with $\beta = 17°$; so, we chose this value. Although this design is guaranteed to be free of cusps, it still produces a sharp curvature. In order to improve this situation, we increase $\bar{e}$ slightly, say, to $\bar{e} = 0.1$, which produced the cam shown in Fig. 4.3.4.

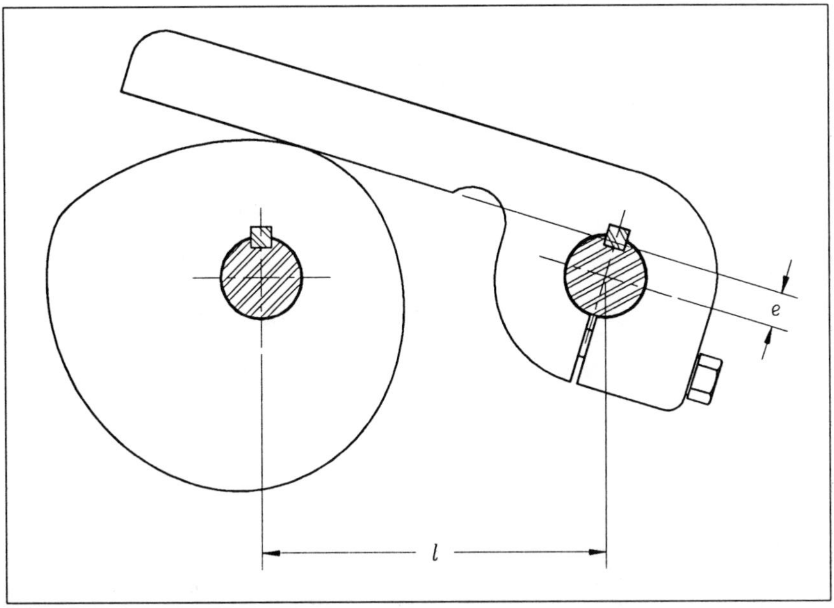

Fig. 4.3.4. Design obtained with $\beta = 17°$ and $\bar{e} = 0.1$.

Chapter 5

OPTIMIZATION OF PLANAR CAM MECHANISMS WITH ROLLER-FOLLOWERS

5.1 Introduction

The approach introduced in Chapter 4 for the optimization of cam mechanisms with flat-face followers is now extended to roller-followers, solid models of which are shown in Figs. 5.1.1 and 5.1.2. The optimization of this type of mechanism has been reported in the literature. In fact, Sermon and Liniecki (1972) treated the problem of cam-size minimization for roller-followers under cycloidal, harmonic, and eighth-order polynomial motions, while Terauchi and El-Shakery (1983) presented the cam-size minimization of roller-follower cam mechanisms, under contact stress constraints. Additionally, Guoxun et al. (1988) introduced a unified approach, applicable to both internal and external indexing cam mechanisms with oscillating roller-followers. Angeles and López-Cajún (1984b, 1988) showed that the cam-size minimization problem for oscillating followers, whether of the flat-face or the roller type, reduces to the solution of a quartic equation. Below we expand the methods proposed in the last two references, in a unified framework, in line with the methods delineated in Chapter 4.

5.2 Translating Knife-Edge and Roller Followers

5.2.1 Knife-Edge Followers

The knife-edge translating follower, shown in Fig. 5.2.1, is not usually found in practice, because of the large contact stresses that it produces (Johnson, 1980).

However, it has theoretical and practical importance for its associated cam profile is the trajectory of the center of the roller follower, to be studied in Subsection 5.2.2. Moreover, if the roller-follower cam is machined with a mill having a radius identical to that of its roller, then the corresponding knife-edge cam profile is identical to the trajectory of the cutter center, the curve needed for the programming of the NC machine tool used to produce the cam disk.

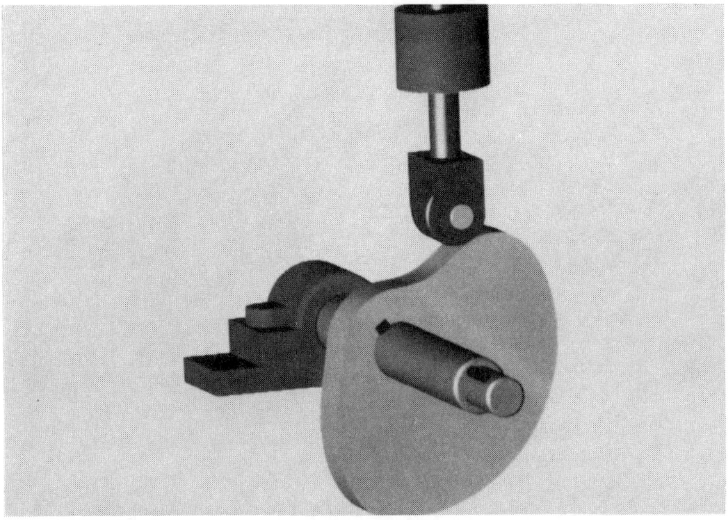

Fig. 5.1.1. Solid model of a planar cam mechanism with a translating follower (produced by Max A. González-Palacios).

It should be pointed out that the purpose of the offset e introduced in this type of cams is, in general, to obtain smaller sizes of cam disks than those obtained with simpler radial cams, for a given maximum pressure angle and displacement program. Thus, the offset is a *design variable* to be determined as explained presently.

In this section, as in the previous chapter, we assume that the displacement program is specified up to a constant c, i.e., the positive-semidefinite function $\sigma(\psi)$ is known, whereas the constant c, a *design variable*, is to be determined. Let the displacement of the follower be denoted by $s(\psi)$, i.e.,

$$s(\psi) = c + \sigma(\psi) \tag{5.2.1}$$

and hence,

$$s'(\psi) = \sigma'(\psi), \qquad s''(\psi) = \sigma''(\psi) \tag{5.2.2}$$

In Fig. 5.2.1, line P is the path of the tip of the follower, whereas lines T and N are the tangent and the normal of the cam profile—the cam profile is given, for example, by its polar equation $\rho = \rho(\theta)$—at the contact point Q. Moreover, lines OF and OC are fixed to the frame and to the cam disk, respectively, e being the offset of the path of the follower tip.

Fig. 5.1.2. Solid model of a planar cam mechanism with an oscillating follower (produced by Max A. González-Palacios).

From the geometry of Fig. 5.2.1, one has

$$s(\psi) = \rho(\theta)\sin(\psi + \theta) \tag{5.2.3a}$$

$$e = \rho(\theta)\cos(\psi + \theta) \tag{5.2.3b}$$

$$\alpha = \psi + \theta + \gamma - \pi \tag{5.2.3c}$$

where α is the pressure angle and γ is obtained from eq. (3.2.4b), namely,

$$\gamma = \tan^{-1}\left[\frac{\rho(\theta)}{\rho'(\theta)}\right] \tag{5.2.3d}$$

The cam profile is readily obtained by combining eqs. (5.2.3a & b). Indeed, these equations lead to the parametric representation of the cam profile, i.e.,

$$\rho(\psi) = \sqrt{s^2(\psi) + e^2} \tag{5.2.4a}$$

$$\theta(\psi) = \tan^{-1}\left[\frac{s(\psi)}{e}\right] - \psi \tag{5.2.4b}$$

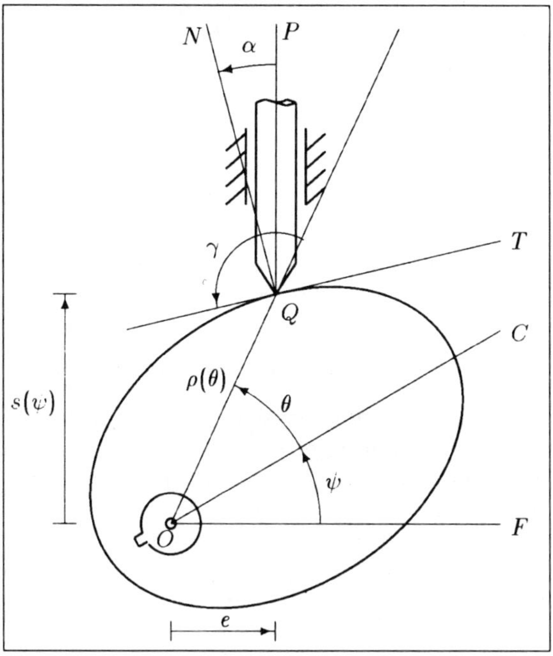

Fig. 5.2.1. Geometry of a cam mechanism with a translating knife-edge follower.

Alternatively, the Cartesian coordinates are computed as explained below. From eqs. (5.2.3*a* & *b*),

$$\rho\,(\sin\psi\cos\theta + \cos\psi\sin\theta) = s \tag{5.2.5a}$$

$$\rho\,(\cos\psi\cos\theta - \sin\psi\sin\theta) = e \tag{5.2.5b}$$

But $\rho\cos\theta = x$ and $\rho\sin\theta = y$ are the Cartesian coordinates of the cam profile in a coordinate frame fixed to the cam. Therefore, eqs. (5.2.5*a* & *b*) can be rewritten as

$$\mathbf{Q}(\psi)\,\mathbf{p}(\psi) = \mathbf{r}(\psi) \tag{5.2.6a}$$

where $\mathbf{Q}(\psi)$, $\mathbf{p}(\psi)$, and $\mathbf{r}(\psi)$ are given by

$$\mathbf{Q}(\psi) = \begin{bmatrix} \cos\psi & -\sin\psi \\ \sin\psi & \cos\psi \end{bmatrix}, \quad \mathbf{p}(\psi) = \begin{bmatrix} x(\psi) \\ y(\psi) \end{bmatrix}, \quad \mathbf{r}(\psi) = \begin{bmatrix} e \\ s(\psi) \end{bmatrix} \tag{5.2.6b}$$

$\mathbf{Q}(\psi)$ being an orthogonal matrix, already introduced in Chapter 4. Note also that $\mathbf{r}(\psi)$ is the position vector of the contact point Q in a coordinate system fixed to the mechanism frame, with origin located at the center of rotation O of the cam disk and coordinate axes oriented along the horizontal and the vertical directions.

From eqs. (5.2.6a), $\mathbf{p}(\psi)$ is obtained as $\mathbf{Q}^T(\psi)\,\mathbf{r}(\psi)$, and hence, $x(\psi)$ and $y(\psi)$ are readily computed, namely,

$$x(\psi) = e\cos\psi + s(\psi)\sin\psi \tag{5.2.7a}$$

$$y(\psi) = -e\sin\psi + s(\psi)\cos\psi \tag{5.2.7b}$$

which are the desired relations in terms of the given displacement $\sigma(\psi)$, as given in eq. (5.2.1), and the parameters c and e to be obtained below.

5.2.2 Cam-Profile Optimization Under a Pressure-Angle Constraint

From the geometry of Fig. 5.2.1, the lowest value of $\rho(\theta)$ is the radius of the base circle b. This is taken as a measure of the cam-disk area and hence, will be the variable to minimize indirectly, under the maximum pressure angle constraint that will be explained presently.

In order to keep the transmitted force of the mechanism within acceptable limits, the pressure angle α is usually bounded properly. Since α can be either positive or negative, we must account for sign changes, i.e.,

$$-\alpha_M \le \alpha \le \alpha_M \tag{5.2.8}$$

a desired value of α being close to 0°, for a value of 90° would render the transmitted force zero, as discussed in Subsection 1.4.1 . A value of α_M that is widely accepted is 30°. Finding the value of ψ at which α attains its extrema is not so simple, for this variable is not readily available. Hence, a function of α that be smooth enough in an interval centered around the origin has to be extremized. A good candidate is $\tan\alpha$, which grows monotonically with α in the interval $-\pi/2 \le \alpha \le \pi/2$. An expression for $\tan\alpha$ is derived below.

From eq. (5.2.3c), clearly,

$$\tan\alpha = \tan(\psi + \theta + \gamma) \tag{5.2.9}$$

The right-hand side of eq. (5.2.9) is computed by resorting to the pertinent trigonometric relation, namely,

$$\tan(\psi + \theta + \gamma) = \frac{\tan(\psi + \theta) + \tan\gamma}{1 - \tan(\psi + \theta)\tan\gamma} \tag{5.2.10}$$

where $\tan(\psi+\theta)$ is obtained by combining eqs. (5.2.3*a* & *b*), i.e.,

$$\tan(\psi+\theta) = s(\psi)/e \tag{5.2.11a}$$

On the other hand, $\tan\gamma$ is obtained from eq.(5.2.3*d*) as

$$\tan\gamma = \frac{\rho(\theta)}{\rho'(\theta)} \tag{5.2.11b}$$

In equation (5.2.11*b*), the computation of $\rho'(\theta)$ is required. This is readily derived, namely,

$$\rho'(\theta) = \rho'(\psi)\,\psi'(\theta) = \frac{\rho'(\psi)}{\theta'(\psi)} \tag{5.2.12}$$

where $\rho'(\psi)$ and $\theta'(\psi)$ are computed in turn as described below. Differentiation of both sides of eqs. (5.2.3*a* & *b*) with respect to ψ leads, respectively, to

$$\rho'(\psi)\sin(\psi+\theta) + \rho(\theta)\,[\,1+\theta'(\psi)\,]\,\cos(\psi+\theta) = s'(\psi) \tag{5.2.13}$$

$$\rho'(\psi)\cos(\psi+\theta) - \rho(\theta)\,[\,1+\theta'(\psi)\,]\sin(\psi+\theta) = 0 \tag{5.2.14}$$

Now, eqs. (5.2.13) and (5.2.14), are cast into vector form, namely,

$$\mathbf{A}\,\mathbf{x} = \mathbf{b} \tag{5.2.15a}$$

where

$$\mathbf{A} = \begin{bmatrix} u & -v\,\rho(\theta) \\ v & u\,\rho(\theta) \end{bmatrix}, \quad \mathbf{x} = \begin{bmatrix} \rho'(\psi) \\ \theta'(\psi) \end{bmatrix}, \quad \mathbf{b} = \begin{bmatrix} v\,\rho(\theta) \\ s'(\psi) - u\,\rho(\theta) \end{bmatrix} \tag{5.2.15b}$$

and the definitions that are shown below:

$$u \equiv \cos(\psi+\theta), \qquad v \equiv \sin(\psi+\theta)$$

From eq. (5.2.15*a*), $\mathbf{x}$ is readily obtained as

$$\mathbf{x} = \mathbf{A}^{-1}\mathbf{b} \tag{5.2.16a}$$

where

$$\mathbf{A}^{-1} = \frac{1}{\rho}\begin{bmatrix} u\,\rho & v\,\rho \\ -v & u \end{bmatrix} \tag{5.2.16b}$$

Therefore,

$$\rho'(\psi) = s'(\psi)\sin(\psi + \theta) \tag{5.2.16c}$$

$$\theta'(\psi) = \frac{s'(\psi)\cos(\psi + \theta) - \rho}{\rho} \tag{5.2.16d}$$

Substitution of eqs. (5.2.16*c* & *d*) into eq. (5.2.12) yields

$$\rho'(\theta) = \frac{\rho\, s'(\psi)\sin(\psi + \theta)}{s'(\psi)\cos(\psi + \theta) - \rho} \tag{5.2.17}$$

Since $\rho\sin(\psi + \theta) = s$ and $\rho\cos(\psi + \theta) = e$, by virtue of eqs. (5.2.3*a* & *b*), we have

$$\rho'(\theta) = \frac{\rho\, s\, s'(\psi)}{e\, s'(\psi) - \rho^2} \tag{5.2.18}$$

Substitution of eq. (5.2.18) into eq. (5.2.11*b*) yields

$$\tan\gamma = \frac{e\, s'(\psi) - \rho^2}{s(\psi)\, s'(\psi)} \tag{5.2.19}$$

while substitution of eq. (5.2.19) and (5.2.11*a*) into eq. (5.2.10) leads to

$$\tan(\psi + \theta + \gamma) = \frac{s^2 s' + e\,(e\, s' - \rho^2)}{\rho^2 s} \tag{5.2.20}$$

the primes indicating differentiation with respect to ψ. From eq. (5.2.4*a*), one obtains ρ^2 which, upon substitution into eq. (5.2.20), yields

$$\tan\alpha = \frac{s' - e}{s} \tag{5.2.21}$$

which is the relation sought.

Now, α will be bounded if $\tan\alpha$ is bounded as well, under the assumption that $-\pi/2 < \alpha < \pi/2$, which is plausible because values of $\alpha = \pm\pi/2$ are ruled out. Thus,

$$-\tan\alpha_M \le \tan\alpha \le \tan\alpha_M \tag{5.2.22}$$

i.e., the conditions below are imposed:

$$\min_{\psi}\{\tan\alpha\} \ge -\alpha_M, \quad \max_{\psi}\{\tan\alpha\} \le \tan\alpha_M \tag{5.2.23}$$

The extrema of $\tan\alpha$, as given by eq. (5.2.21), are now found by zeroing the derivative of $\tan\alpha$ with respect to ψ, namely,

$$\frac{d\tan\alpha}{d\psi}=0 \tag{5.2.24a}$$

which yields

$$s''s-s'(s'-e)=0 \tag{5.2.24b}$$

or

$$\frac{s'-e}{s}=\frac{s''}{s'} \tag{5.2.24c}$$

Let ψ^* be the value of ψ, as yet undetermined, producing a stationary value of $\tan\alpha$ that equals the maximum allowable value, i.e., either $\tan\alpha_M$ or $-\tan\alpha_M$. Thus,

$$\frac{s'(\psi^*)-e}{s(\psi^*)}=\tan\alpha_M \quad \text{or} \quad \frac{s'(\psi^*)-e}{s(\psi^*)}=-\tan\alpha_M \tag{5.2.25}$$

Recalling eqs. (5.2.2), namely, $s'(\psi)=\sigma'(\psi)$ and $s''(\psi)=\sigma''(\psi)$, eqs. (5.2.24c) and (5.2.25) lead to

$$\frac{\sigma''(\psi^*)}{\sigma'(\psi^*)}=\tan\alpha_M \quad \text{or} \quad \frac{\sigma''(\psi^*)}{\sigma'(\psi^*)}=-\tan\alpha_M \tag{5.2.26a}$$

which can be rewritten as

$$\sigma''(\psi^*)=\tan\alpha_M\,\sigma'(\psi^*) \quad \text{or} \quad \sigma''(\psi^*)=-\tan\alpha_M\,\sigma'(\psi^*) \tag{5.2.26b}$$

The graphical solution of eqs. (5.2.26b) is shown in Fig. 5.2.2, for $\sigma(\psi)$ defined as a cycloidal motion, $\alpha_M=30°$, and a duration of the rise phase $\Delta\psi=4\pi/5$ rad $=144°$. From that figure, *four* solutions to the two foregoing equations are determined, namely, $\psi^*=\psi_a$, $\psi^*=\psi_b$, and the two intermediate ones, labeled ψ_1 and ψ_2. Let

$$\sigma_i\equiv\sigma(\psi_i),\quad \sigma_i'\equiv\sigma'(\psi_i),\quad \sigma_i''\equiv\sigma''(\psi_i),\quad \text{for } i=1,2$$

Now, the set of values along which condition (5.2.26b) holds during the rise phase is defined upon substitution of σ_1 and σ_2 into eqs. (5.2.25), together with the condition

$$c>0 \tag{5.2.27}$$

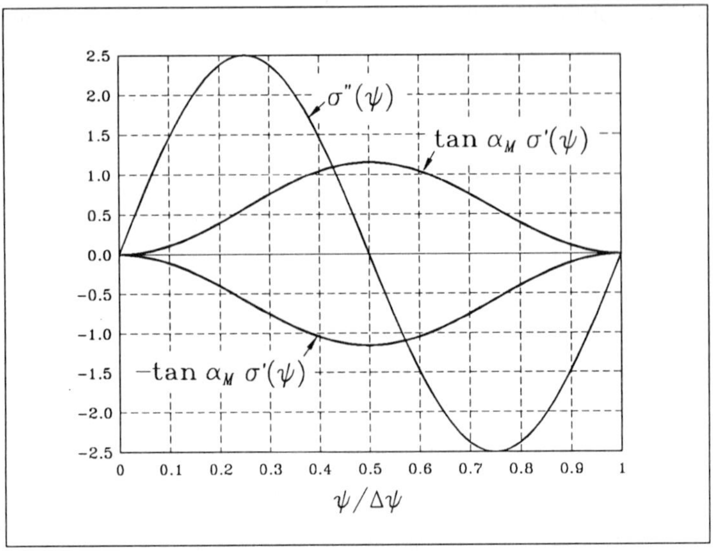

Fig. 5.2.2. Intersection of $\sigma''(\psi)$ with both $\tan\alpha_M\,\sigma'(\psi)$ and $-\tan\alpha_M\,\sigma'(\psi)$

Thus, eqs. (5.2.25) are written at the two intermediate extrema of $\tan\alpha$ found from Fig. 5.2.2, i.e.,

$$\sigma_1' - e = \tan\alpha_M\,(\sigma_1 + c) \qquad (5.2.28a)$$

and

$$\sigma_2' - e = -\tan\alpha_M\,(\sigma_2 + c) \qquad (5.2.28b)$$

which are now two equations in two unknowns, namely, the design variables c and e. These equations can be cast into a standard form, namely,

$$e = -\tan\alpha_M\,c + k_1 \qquad (5.2.28c)$$

$$e = \tan\alpha_M\,c + k_2 \qquad (5.2.28d)$$

where k_1 and k_2 are defined as

$$k_1 \equiv \sigma_1' - \tan\alpha_M\,\sigma_1, \qquad k_2 \equiv \sigma_2' + \tan\alpha_M\,\sigma_2 \qquad (5.2.28e)$$

For a *symmetric* displacement program like the cycloidal motion, due to the symmetries involved, $\sigma_1' = \sigma_2'$ and $\sigma_2 > \sigma_1$. Thus, $k_2 > k_1 > 0$ and hence, the plots of eqs. (5.2.28*c* & *d*) take on the form shown in Fig. 5.2.3. In that figure, lines $\mathcal{L}_1$ and $\mathcal{L}_2$ make an angle α_M and $-\alpha_M$ with the c axis, respectively, while $k_1 = \overline{OB}$, and $k_2 = \overline{OD}$. Moreover, since the pressure angle also attains stationary values at the lowest follower position, as discussed earlier, the relation below, arising from Fig. 5.2.4, can also be imposed: that is, for any set of $(c,\ e)$ values yielding a pressure angle of α_M at the lower dwell, we have

$$e = c \tan \alpha_M \tag{5.2.29}$$

Clearly, eq. (5.2.29) represents a line parallel to the line described by eq. (5.2.28d) and hence, there will be no intersection between them. Therefore, eqs. (5.2.28c) and (5.2.29) are to be considered as the design equations for finding c and e. Equation (5.2.29) is also plotted in Fig 5.2.3. From that figure, it is clear that point P of coordinates $(c_0,\ e_0)$ simultaneously satisfies relations (5.2.27), (5.2.28c), and (5.2.29). Thus, coordinates c_0 and e_0 represent the optimum design sought. These coordinates are given below as

$$c_0 = \frac{k_1}{2 \tan \alpha_M}, \qquad e_0 = \frac{k_1}{2} \tag{5.2.30a}$$

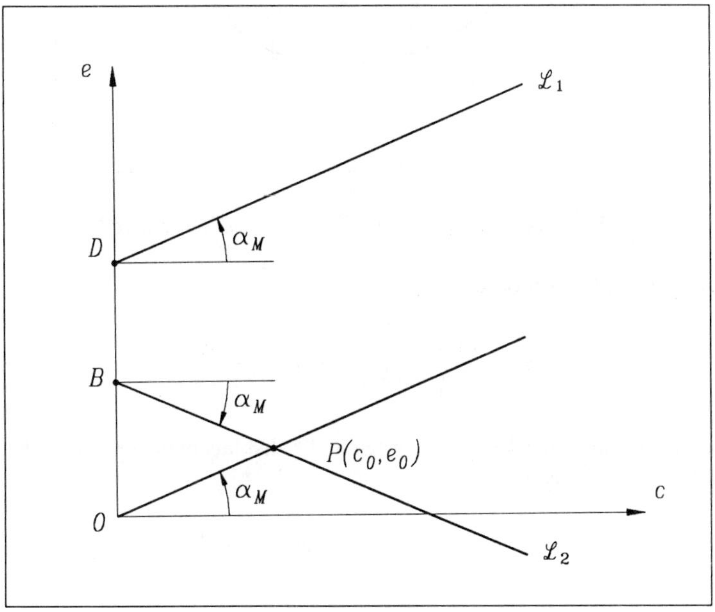

Fig. 5.2.3. Plots of equations (5.2.28c & d) and (5.2.29).

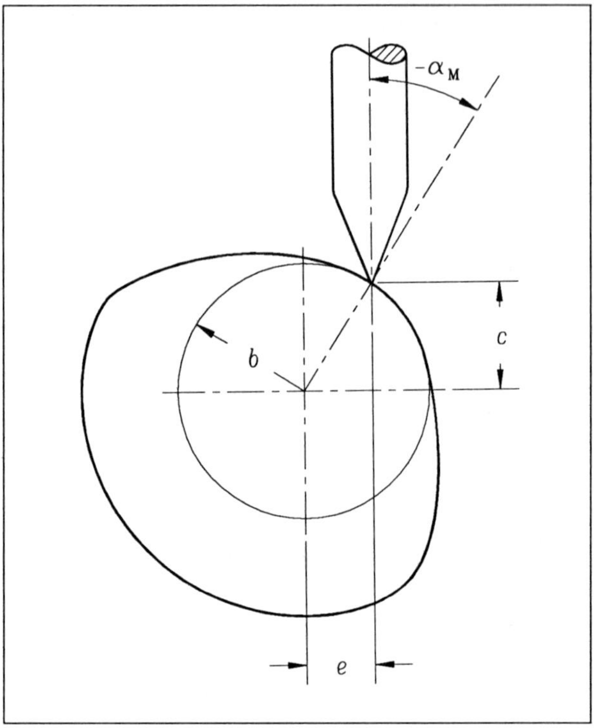

Fig. 5.2.4. Knife-edge follower at its lower dwell.

from which the optimum base radius b_0 is given by

$$b_0 = \frac{k_1}{2}\sqrt{\frac{1}{\tan^2 \alpha_M} + 1} = \frac{k_1}{2 \sin \alpha_M} \tag{5.2.30b}$$

If a normal motion is used for synthesizing the displacement program, k_1, as given by eq. (5.2.28*e*), reduces to

$$\begin{aligned} k_1 &= \frac{h}{\Delta\psi}\,\tau'(x^*) - h \tan \alpha_M \tau(x^*) = \\ &= \frac{h}{\Delta\psi}\,[\,\tau'(x^*) - K\,\tau(x^*)\,] \end{aligned} \tag{5.2.30c}$$

where

$$K \equiv \Delta\psi \, \tan \alpha_M \tag{5.2.30d}$$

The above derivation holds for any displacement program, except for the case of constant velocity, which, nevertheless, is unlikely to occur in practice for the reasons given in Section 2.3.

We have been assumed so far that ψ^* is known. In fact, from the foregoing discussion, we are interested in the values of ψ satisfying the first of eqs. (5.2.26*b*), which correspond to those that zero the function

$$f(\psi) = \sigma''(\psi) - \tan\alpha_M\,\sigma'(\psi), \qquad \psi_a \le \psi \le \psi_b \tag{5.2.31a}$$

where $\psi_b - \psi_a$ represents the cam angle of rotation during the rise phase, i.e., $\Delta\psi$. Finding the value ψ^* of ψ, for which $f(\psi)$ of eq. (5.2.31*a*) vanishes can be done as explained below. If graphics software is available, eq. (5.2.31*a*) can be displayed, and, from there, one can infer visually the root, i.e., ψ^*. This can be done by plotting the two functions $\sigma''(\psi)$ and $\tan\alpha_M\,\sigma'(\psi)$ on a graphics monitor, the desired value ψ^* being found at the intersection of the two plots. The numerical value of ψ^* can be obtained in turn using a *mouse*—see Chapter 9—as an aid to point to the said intersection. Otherwise, one will have to resort to a root-finding algorithm such as *bisection*, *tangent*, *Fibonacci*, or *golden section* (Forsythe et al., 1977).

If the displacement program s is prescribed based on a normal motion, then, finding the value of $\psi = \psi^*$, so that $f(\psi^*) = 0$, is an easy task. Indeed, upon substitution of eqs. (2.2.2) into eq. (5.2.31*a*), $f(\psi)$ is obtained as

$$f(\psi) = \frac{h}{(\Delta\psi)^2}\,\tau''(x) - \frac{h}{\Delta\psi}\,\tan\alpha_M\,\tau'(x) \tag{5.2.31b}$$

Furthermore, multiplying both sides of eq. (5.2.31*b*) by $(\Delta\psi)^2/h$,

$$g(x) \equiv \frac{(\Delta\psi)^2}{h}\,f(\psi) = \tau''(x) - K\,\tau'(x), \qquad 0 \le x \le 1 \tag{5.2.32a}$$

where K was defined in eq. (5.2.30*d*).

Note that, if a 3-4-5 polynomial motion is used, then finding the roots of $g(x)$, as given in eq. (5.2.32*a*), amounts to solving a quartic equation. However, since both $\tau'(x)$ and $\tau''(x)$ vanish at $x = 0$ and $x = 1$, $x\,(x-1)$ is a factor of $g(x)$ and hence, the quartic equation reduces to a quadratic equation. In fact, upon substitution of $\tau'(x)$ and $\tau''(x)$, as given in eqs. (2.3.19) into $g(x)$ as given in eq. (5.2.32*a*), and zeroing $g(x)$ next, leads to

$$60\,x\,(x-1)\,(2\,x-1) - 30\,K\,x^2\,(x-1)^2 = 0$$

which thus reduces to

$$2\,(2\,x-1) - K\,x\,(x-1) = 0$$

or

$$K\,x^2 - (4+K)\,x + 2 = 0$$

whose *significant* root is given below:

$$x^* = \frac{4+K-\sqrt{16+K^2}}{2K} \tag{5.2.32b}$$

the other root being discarded, for x^* must lie in the interval $[\,0,\ 1\,]$. Indeed, let us consider the right triangle shown in Fig. 5.2.5 and assume that $K < 4$. This assumption is plausible, for α_M is usually given as 30° and never greater than 45°, while $\Delta\psi$ is seldom greater than π. For example, for a value of $\alpha_M = 45°$, $\Delta\psi$ must be greater than 229.2° in order to produce a value of K greater than 4.

From the triangle in Fig. 5.2.5, then, the two inequalities given below must hold:

$$4+K > \sqrt{16+K^2}$$

$$4-K < \sqrt{16+K^2}$$

The first of the two foregoing inequalities guarantees that x^* as given by eq. (5.2.32b), is positive, while the second inequality can be rewritten as

$$4+K-2\,K < \sqrt{16+K^2}$$

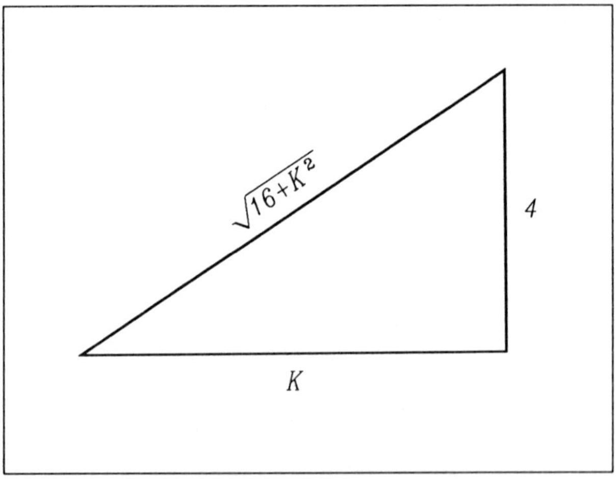

Fig. 5.2.5. Geometric interpretation of the solution x^* appearing in eq. (5.2.32b).

and hence,

$$4 + K - \sqrt{16 + K^2} < 2K$$

thereby showing that x^*, as given by eq. (5.2.32b), is smaller than unity.

Now, upon substituting eq. (5.2.32b) into (5.2.30c) for the 3-4-5 polynomial motion and using computer algebra (Wolfram, 1988), we obtain

$$\eta(K) \equiv \tau' - K\tau = \frac{192}{K^4}\left(4 - \sqrt{16 + K^2}\right) +$$

$$+ \frac{4}{K^2}\left(10 - \sqrt{16 + K^2}\right) - \frac{1}{2}\left(K - \sqrt{16 + K^2}\right)$$

Substitution of the above expression into eq. (5.2.30a) yields the optimum parameters c_0 and e_0 as

$$c_0 = \frac{h\,\eta(K)}{2K}$$

and

$$e_0 = \frac{h\,\eta(K)}{2\Delta\psi}$$

Thus, the optimum base radius, as given by eq. (5.2.30b), reduces to

$$\frac{b_0}{h} = \frac{1}{2K\cos\alpha_M}\left[\frac{192}{K^4}(4 - S) + \frac{4}{K^2}(10 - S) - \frac{1}{2}(K - S)\right] \tag{5.2.32c}$$

with S defined as

$$S \equiv \sqrt{16 + K^2} \tag{5.2.32d}$$

Equation (5.2.32c) allows us to determine the optimum base radius explicitly.

By the same token, if a cycloidal motion is used, then $\tau'(x)$ and $\tau''(x)$ vanish at $x = 0$ and $x = 1$ and then we are interested in the intermediate roots. However, $\tau'(x)$ is linear in $\cos 2\pi x$ and $\tau''(x)$ is linear in $\sin 2\pi x$. As a consequence, upon introduction of the trigonometric identities

$$\cos 2\pi x \equiv \frac{1 - t^2}{1 + t^2}, \quad \sin 2\pi x \equiv \frac{2t}{1 + t^2}, \qquad t \equiv \tan \pi x$$

finding x^* that zeroes $g(x)$ as given by eq. (5.2.32a) reduces to solving a quadratic equation in t that, nevertheless, simplifies to a linear equation. Indeed, if we substitute $\tau'(x)$ and $\tau''(x)$, as given in eqs. (2.3.6b & c), into $g(x)$, as appearing in eq. (5.2.32a), and then zero the arising relation, we obtain

$$2\pi \sin 2\pi x - K(1 - \cos 2\pi x) = 0$$

Next, we introduce the foregoing trigonometric identities into the last equation, thereby deriving a quadratic equation, namely,

$$-2\,K\,t^2 + 4\,\pi\,t = 0$$

Since we are interested in a nonvanishing solution, we assume $t \neq 0$ and hence, both sides of the above equation can be divided by t, which yields

$$x^* = \frac{1}{\pi}\,\tan^{-1}\left(\frac{2\pi}{K}\right) \tag{5.2.32e}$$

Substitution of eq. (5.2.32e) into (5.2.30a) yields

$$k_1 = \frac{h}{\Delta\psi}\left[2 - \frac{K}{\pi}\,\tan^{-1}\left(\frac{2\pi}{K}\right)\right]$$

and the optimum parameters are computed as

$$c_0 = \frac{h}{2K}\left[2 - \frac{K}{\pi}\,\tan^{-1}\left(\frac{2\pi}{K}\right)\right]$$

$$e_0 = \frac{h}{2\Delta\psi}\left[2 - \frac{K}{\pi}\,\tan^{-1}\left(\frac{2\pi}{K}\right)\right]$$

the optimum base radius being

$$\frac{b_0}{h} = \frac{1}{2\,K\,\cos\alpha_M}\left[2 - \frac{K}{\pi}\,\tan^{-1}\left(\frac{2\pi}{K}\right)\right] \tag{5.2.32f}$$

Note that, from eqs. (5.2.32c & f), the ratio b_0/h can be expressed as

$$\frac{b_0}{h} = \frac{F(K)}{2\,K\,\cos\alpha_M} \tag{5.2.32g}$$

where

$$F(K) \equiv \begin{cases} \frac{192}{K^4}\,(4 - S) + \frac{4}{K^2}\,(10 - S) - \frac{1}{2}\,(K - S)\,, & \text{for 3-4-5 polynomial motion} \\ 2 - \frac{K}{\pi}\,\tan^{-1}\left(\frac{2\pi}{K}\right), & \text{for cycloidal motion} \end{cases}$$

From the above relations, it should be possible to show that a 3-4-5 polynomial motion produces smaller optimum cams than cycloidal motions.

In general, once the value of x^* and, correspondingly, ψ^* has been determined, one can proceed to determining the values of $\sigma(\psi^*)$, $\sigma'(\psi^*)$, and $\sigma''(\psi^*)$ and, hence, the values of c_0 and e_0, which will determine in turn the minimum radius of the base circle, b_0. From Fig. 5.2.4, for $b = b_0$, $c = c_0$, and $e = e_0$,

$$b_0 = \sqrt{c_0^2 + e_0^2} \tag{5.2.33a}$$

where

$$c_0 = \frac{1}{2}\overline{OA}, \qquad e_0 = \frac{1}{2}\overline{OB} \tag{5.2.33b}$$

while

$$\overline{OA} = \frac{k_1}{\tan\alpha_M} = \frac{\sigma'(\psi^*)}{\tan\alpha_M} - \sigma(\psi^*) \tag{5.2.33c}$$

and

$$\overline{OB} = k_1 = \sigma'(\psi^*) - \tan\alpha_M\,\sigma(\psi^*) \tag{5.2.33d}$$

Upon substitution of eqs. (5.2.33*b*–*d*) into eq. (5.2.33*a*), we obtain eq. (5.2.30*b*), reproduced here for quick reference

$$b_0 = \frac{k_1}{2\sin\alpha_M} \tag{5.2.33e}$$

thereby completing the optimum design. Note that, if $\alpha_M = 30°$, a rather frequent value, then b_0 is identical to k_1. A fully-detailed example following the aforementioned methodology is given below.

5.2.3 Example 5.2.1: Minimum-Size Cam Profile for a Knife-Edge Follower

Obtain the cam profile of the translating knife-edge follower cam mechanism producing the displacement program given in Table 5.2.1, for a maximum pressure angle of 30°. Use

i) a 3-4-5 polynomial motion

ii) a cycloidal motion

Table 5.2.1: Prescribed angular-displacement program.

Phase	Percentage of cycle	Displacement (mm)
D_1 (dwell)	10	0
R_1 (rise)	40	+50
D_2 (dwell)	25	0
R_2 (return)	25	−50

Solution:

We will base our design on the rise phase. From the given data, one has

$$h = 50, \quad \psi_a = 0.2\,\pi, \quad \psi_b = \pi, \quad \psi_c = 1.5\,\pi,$$

Thus,

$$\Delta\psi \equiv \psi_b - \psi_a = 0.8\,\pi = 2.5133 \text{ rad } (= 144^\circ)$$

and hence,

$$K = \Delta\psi \tan\alpha_M = 0.8\,\pi \tan\frac{\pi}{6} = 1.451$$

i) 3-4-5 Polynomial motion

With the value of K derived above, we can readily compute, from eq. (5.2.32*b*),

$$x^* = 0.4121$$

and hence,

$$\psi^* = \psi_a + x^*\,\Delta\psi = 0.2\,\pi + 0.4121 \times 2.5133 = 1.664 \text{ rad} = 95.34^\circ$$

Now we can calculate k_1 as given by eq. (5.2.30*c*), namely,

$$k_1 = \frac{h}{\Delta\psi}\left[30\,x^{*2}(x^* - 1)^2 - K\,x^{*3}(10 - 15\,x^* + 6\,x^{*2})\right] =$$

$$= \frac{50}{2.5133}\left[30\,x^{*2}(x^* - 1)^2 - 1.451\,x^{*3}(10 - 15\,x^* + 6\,x^{*2})\right] =$$

$$= 25.26 \text{ mm}$$

The optimum design can now be computed from eqs. (5.2.30*a* & *b*), namely,

$$c_0 = \frac{25.26}{2 \times 0.5773} = 21.88 \text{ mm}$$

$$e_0 = \frac{25.26}{2} = 12.63 \text{ mm}$$

and hence, the optimum base radius b_0 is given as

$$b_0 = \sqrt{21.88^2 + 12.63^2} = 25.26 \text{ mm}$$

which shows that, in this case, as anticipated by eq. (5.2.30*b*), $b_0 = k$, for $\alpha_M = 30°$. The cam profile for this design is shown in Fig. 5.2.6.

ii) Cycloidal motion

The displacement program is now generated using the normal cycloidal motion, as explained in Subsection 2.5.1. Then, the value x^*, is obtained using eq. (5.2.32*c*), namely,

$$x^* = \frac{1}{\pi} \tan^{-1}\left(\frac{2\pi}{K}\right)$$

which yields $x^* = 0.42776$. This value can be verified visually by plotting function $g(x)$ of eq. (5.2.32) as shown in Fig. 5.2.7. The corresponding value of ψ, derived from eqs. (2.4.26*b*), for $x = x^*$ and $\psi = \psi^*$, is computed below

$$\psi^* = \psi_a + x^* \Delta\psi = 0.2\,\pi + 0.4278 \times 2.5133 = 1.7035 \text{ rad} = 97.60°$$

i.e., the pressure angle attains a maximum value during the rise phase when ψ is 97.60°.

Next, the optimum parameters are obtained using eqs. (5.2.33*a–d*), where k_1 is computed using eq. (5.2.30*c*), namely,

$$k_1 = \frac{h}{\Delta\psi}\left[\tau'(x^*) - K\tau(x^*)\right]$$

The above formula yields $k_1 = 27.44$ mm and, correspondingly, $e_0 = 13.72$ mm and $c_0 = 23.76$ mm. With these values, the optimum radius of the base circle is calculated using eq. (5.2.33*a*), which yields $b_0 = 27.44$ mm. Next, the Cartesian coordinates of the cam profile are obtained using eqs. (5.2.7*a* & *b*). The cam profile is shown in Fig. 5.2.8.

Alternatively, in both cases, the optimum base radius can be computed using eq. (5.2.32*g*), which yields the same values than above. Note that the 3-4-5 polynomial motion produced a cam with a base radius of 92.05% of that produced with a cycloidal motion. As a consequence, the former cam will have an area that is roughly 85% that of the latter. In industrial production of big batches of cams, or in cases where weight is crucial, these savings may be quite attractive.

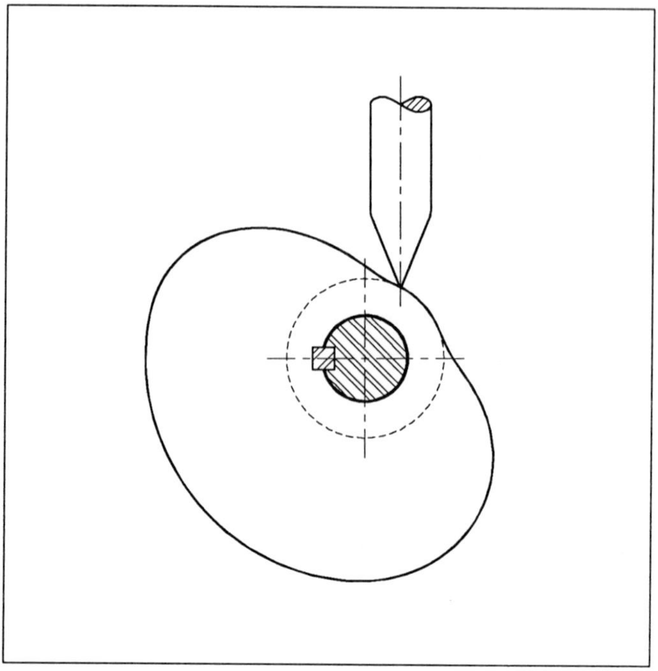

Fig. 5.2.6. Cam profile for the 3-4-5 polynomial motion.

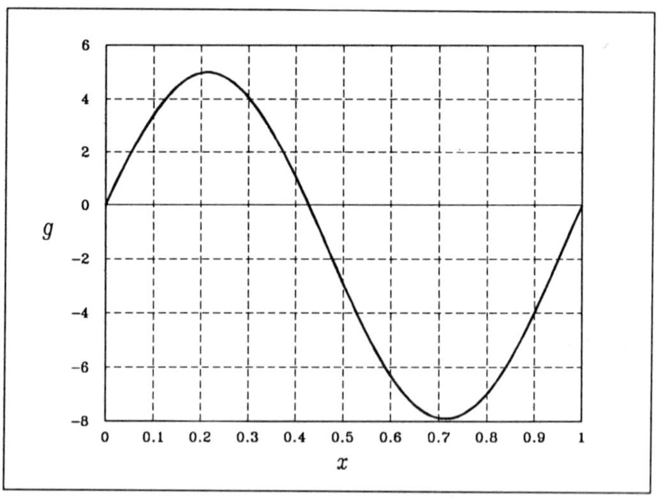

Fig. 5.2.7. Plot of function $g(x)$ of eq. (5.2.32a) for normal cycloidal motion.

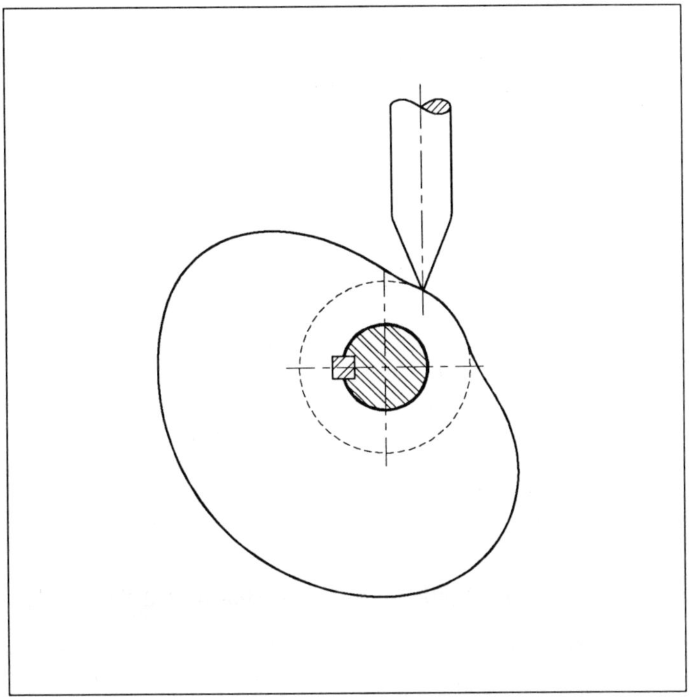

Fig. 5.2.8. Cam profile for a cycloidal motion.

5.2.4 Radial Followers

If the offset e is set equal to zero, eq. (5.2.21) yields

$$\tan \alpha = \frac{s'(\psi)}{s(\psi)} \geq 0, \quad \text{for the rise phase} \tag{5.2.34}$$

where the inequality follows by virtue of the sign of $s'(\psi)$, which is nonnegative in the rise phase, given that, by definition, $s(\psi)$ is positive. Hence, of the two relations (5.2.28a & b), only the former holds, which thus yields

$$c = b_r = \frac{\sigma_1'}{\tan \alpha_M} - \sigma_1$$

or

$$b_r = \frac{k_1}{\tan \alpha_m} \tag{5.2.35}$$

with k_1 defined in eq. (5.2.28*e*).

The base radii for the radial follower and the offset follower are related as discussed presently. From eq. (5.2.30*b*), b_0 can be expressed as

$$b_0 = \frac{k_1}{2\sin\alpha_M} \equiv \frac{1}{2\cos\alpha_M}\frac{k_1}{\tan\alpha_M}$$

and hence,

$$\frac{b_0}{b_r} = \frac{1}{2\cos\alpha_M} \tag{5.2.36}$$

That is,

$$\frac{b_0}{b_r}\begin{cases} < 1, & \text{for } \alpha_M < 60^\circ; \\ \geq 1, & \text{for } \alpha_M \geq 60^\circ \end{cases}$$

Therefore, for acceptable values of α_M, usually under 45°, $b_0 < b_r$, and hence, the presence of an offset yields smaller cams, a quite practical result.

5.2.5 Example 5.2.2: Cam Profile for a Translating Radial Knife-Edge Follower

For the data given in Example 5.2.1, design the cam profile for a radial follower.

Solution:

In order to fully exploit the results of Example 5.2.1, we will use eq. (5.2.36) in the form

$$b_r = 2\, b_0 \cos\alpha_M$$

Thus, for each of the two normal motions treated in Example 5.2.1, we have

i) 3-4-5 Polynomial motion:

$$b_r = 2 \times 25.26 \times 0.8660 = 43.75 \text{ mm}$$

ii) Cycloidal motion:

$$b_r = 2 \times 27.44 \times 0.8660 = 47.53 \text{ mm}$$

The resulting cam profiles are shown in Figs. 5.2.9*a* & *b*.

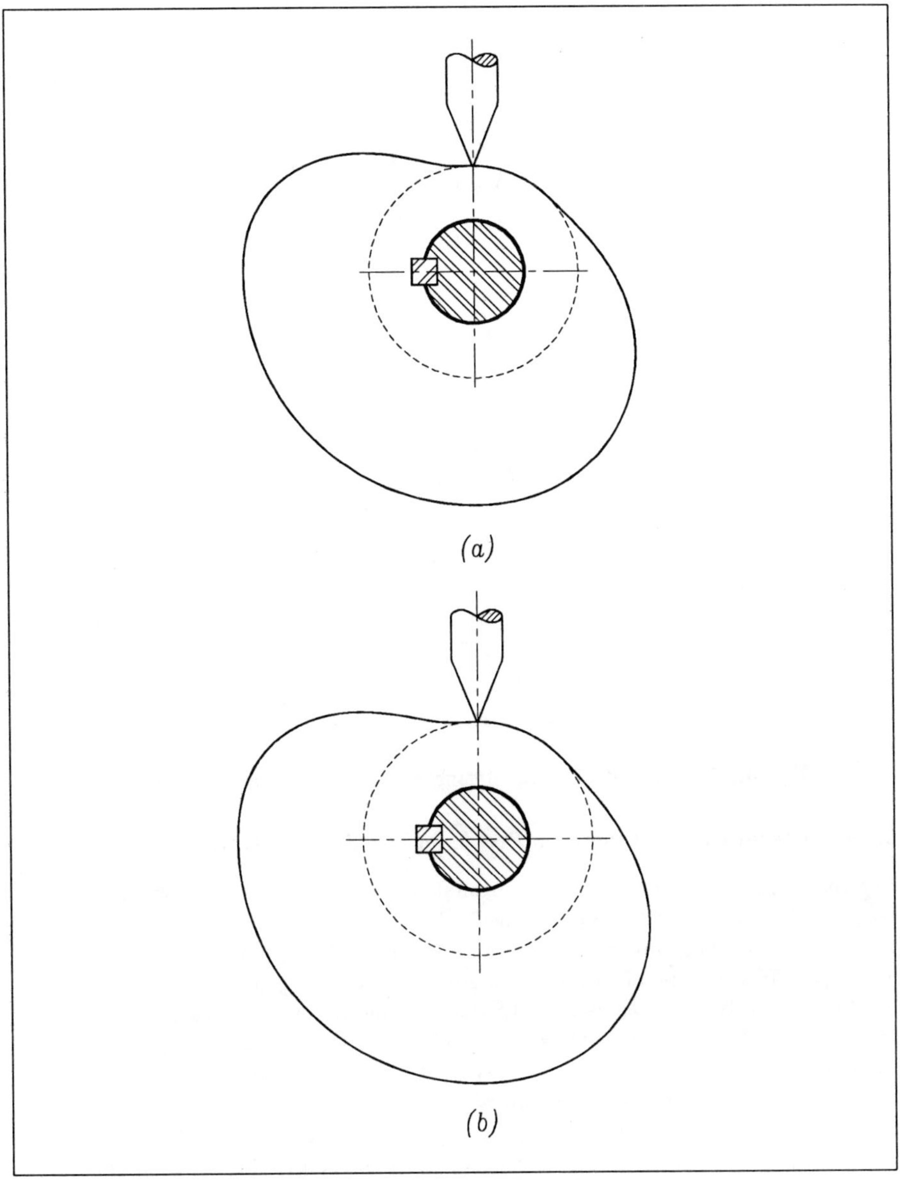

Fig. 5.2.9. Cam profile for a translating radial knife-edge follower, for *a*) 3-4-5 polynomial motion, and *b*) cycloidal motion.

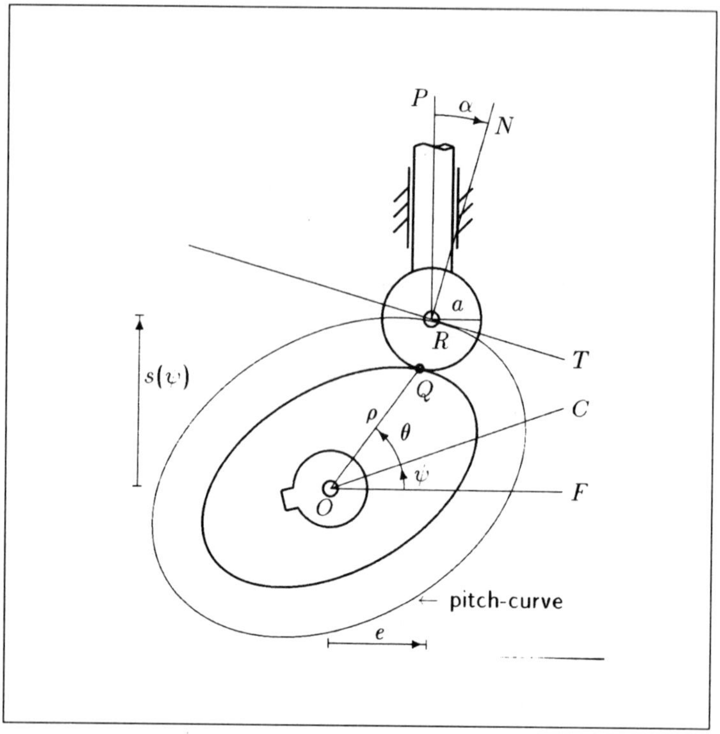

Fig. 5.2.10. Layout of a translating roller-follower cam mechanism.

5.2.6 Determination of the Cam Profile for the Roller-Follower

In Subsection 5.2.1, the Cartesian coordinates describing the cam profile for a translating knife-edge follower were obtained. Based on those results, the corresponding methodology for determining the cam profile for a roller follower is explained below. The layout of this type of cam mechanism is shown in Fig. 5.2.10.

In Fig. 5.2.10, line OF is fixed to the machine frame, whereas line OC is fixed to the cam. Moreover, the radius of the roller and the follower offset are indicated by a and e, respectively, whereas ρ—segment $\overline{OQ}$—and θ are the polar coordinates describing the cam profile. The cam angle of rotation is represented in turn by ψ and the follower output by $s(\psi)$, while the pressure angle is denoted by α, as before.

Let $\mathbf{e}_t$, $\mathbf{e}_n$, and r be the unit tangent and normal vectors of the cam profile and its radius of curvature, respectively. As discussed in Subsection 3.2.2, r can be positive or negative, depending upon the relative location of the center of curvature of the cam profile and the cam center of rotation with respect to the cam profile. Clearly, $\mathbf{e}_t$, $\mathbf{e}_n$, and r are dependent upon the angle of rotation ψ. Now, specifying a value a

of the roller radius, the profile of the corresponding cam can be obtained from the relation

$$\mathbf{r}_r = \mathbf{r}_k + a\,\mathbf{e}_n \tag{5.2.37}$$

where $\mathbf{r}_r$ and $\mathbf{r}_k$, shown in Fig. 5.2.11, are the position vectors of *corresponding* points—points sharing a common normal—on the cam profile for the roller and for the knife-edge followers, respectively. Here, in order to avoid the phenomenon known as *undercutting*, a is selected as the maximum allowable value of the radius of the roller. This phenomenon is further discussed below—see also Subsection 1.4.3. On the other hand, in order to avoid large values of the contact stress, it is also necessary to determine the minimum absolute value of the radius of curvature of the cam profile, r_m, for the roller. However, the radius of curvature is infinite in sections where the cam profile is a straight line and also at inflection points—when the cam profile changes from convex to concave, or conversely. On the contrary, the curvature remains bounded and vanishes at inflection points and at linear segments of the profile. Thus, in the discussion below we will be concerned with finding the maximum absolute value of the curvature, rather than the minimum absolute value of the radius of curvature.

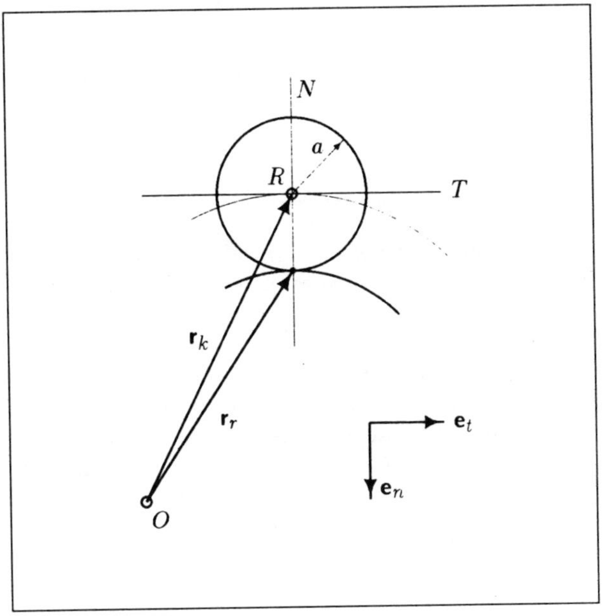

Fig. 5.2.11. Geometry of a cam profile and path of the roller center R.

Let κ_k and κ_r be the curvature of the cam profile for the knife-edge and for the roller-followers, respectively. Since the tangents of both profiles at corresponding points are parallel, one readily obtains

$$\kappa_r = \frac{\kappa_k}{1 - a\kappa_k} \tag{5.2.38}$$

If the denominator of the right-hand side of eq. (5.2.38) vanishes, the curvature of the roller-follower cam profile, κ_r, will become infinite. This means that, at values of ψ where that denominator vanishes, the said cam profile has a cusp, a defect known as *undercutting*. Moreover, both the *pitch curve*, shown in Fig. 5.2.10, and the roller-follower cam profile should have, at corresponding points, curvatures with the same sign. In light of eq. (5.2.38), this is attained if, and only if,

$$1 - a\,\kappa_k > 0, \qquad 0 \leq \psi \leq 2\pi \tag{5.2.39a}$$

Relation (5.2.39a) holds, in turn, if

$$a > \frac{1}{\kappa_M}, \qquad \kappa_M \equiv \max_{\psi}\{|\,\kappa_k(\psi)\,|\} \tag{5.2.39b}$$

Alternatively, if a is specified as a given fraction ν of r_m, the minimum absolute value of the radius of curvature, then

$$a = \nu\, r_m = \frac{\nu}{\kappa_M}, \qquad 0 < \nu < 1 \tag{5.2.39c}$$

The maximum absolute value of κ_k, κ_M, is now determined. This can be done, again, either interactively with a graphics monitor, plotting software, and a *mouse*, or numerically, as explained presently. In order to calculate κ_M numerically, $\kappa_k'(\psi)$ is zeroed and the values of the curvature κ_k at stationary points are computed. It can be readily shown that κ_k attains stationary values at the dwell phases. However, none of these is, in general, a global extremum, and hence, the global maximum is sought. Note that, while we were concerned only with the rise phase in designing for a maximum value of pressure angle, here we have to consider both the rise and the return phases. Moreover, the curvature κ_k is computed below using eq. (3.2.16), with sgn $[\,\lambda'(q)\,] = -1$ as explained in Subsection 4.2.4, namely,

$$\kappa = \frac{\mathbf{p}'^T(\psi)\,\mathbf{E}\,\mathbf{p}''(\psi)}{||\,\mathbf{p}'(\psi)\,||^3} \tag{5.2.40a}$$

where the parameter q has been replaced by ψ, while $\mathbf{p}'$ and $\mathbf{p}''$ were already obtained in Subsection 4.2.4, namely,

$$\mathbf{p}'(\psi) = \mathbf{Q}^T(\psi)\,[\,\mathbf{r}'(\psi) - \mathbf{E}\,\mathbf{r}(\psi)\,] \tag{5.2.40b}$$

$$\mathbf{p}''(\psi) = \mathbf{Q}^T(\psi)\,\left[\,\mathbf{r}''(\psi) - 2\,\mathbf{E}\,\mathbf{r}'(\psi) - \mathbf{r}(\psi)\,\right] \tag{5.2.40c}$$

For the case at hand, $\mathbf{r}$ is obtained from eq. (5.2.6b), its corresponding derivatives being shown below, along with $\mathbf{r}$ itself, for quick reference:

$$\mathbf{r}(\psi) = \begin{bmatrix} e \\ s(\psi) \end{bmatrix}, \quad \mathbf{r}'(\psi) = \begin{bmatrix} 0 \\ s'(\psi) \end{bmatrix}, \quad \mathbf{r}''(\psi) = \begin{bmatrix} 0 \\ s''(\psi) \end{bmatrix} \tag{5.2.41}$$

Upon substitution of $\mathbf{r}$, $\mathbf{r}'$, and $\mathbf{r}''$ as given by (5.2.41) into eqs. (5.2.40a–c) we obtain, after simplifying the ensuing relations,

$$\kappa_k = \frac{N}{D} \tag{5.2.42a}$$

where

$$N \equiv s\,(s - s'') + (s' - e)\,(2s' - e) \tag{5.2.42b}$$

and

$$D \equiv [\,s^2 + (s' - e)^2\,]^{3/2} \tag{5.2.42c}$$

Now, the stationary points of κ_k are found from the normality condition as explained below. Differentiation of both sides of eq. (5.2.42a) with respect to ψ yields

$$\kappa_k'(\psi) = \frac{1}{D}\,[\,N'(\psi) - \kappa_k\,D'(\psi)\,] = 0 \tag{5.2.43}$$

From eq. (5.2.42b & c), $N'(\psi)$ and $D'(\psi)$ are readily derived, namely,

$$N'(\psi) = s\,(\,2\,s' - s'''\,) + 3\,s''\,(\,s' - e\,) \tag{5.2.44a}$$

$$D'(\psi) = 3\,[\,ss' + s''(s' - e)\,]\,[\,s^2 + (s' - e)^2\,]^{1/2} \tag{5.2.44b}$$

Thus, eq. (5.2.43) leads to

$$s\,(2\,s' - s''') + 3\,s''\,(s' - e)$$
$$- 3\,\kappa_k[\,s\,s' + s''\,(s' - e)\,]\,[\,s^2 + (s' - e)^2\,]^{1/2} = 0 \tag{5.2.44c}$$

Upon substitution of κ_k as given by eqs. (5.2.40a) and (5.2.42a & b) into eq. (5.2.44c), one obtains, after simplifications,

$$H(\psi) \equiv [\, s^2 + (s' - e)^2 \,] [\, s\,(2\,s' - s''') + 3\,s''\,(s' - e)\,] -$$

$$- 3\,[\, s\,s' + s''\,(s' - e)\,] [\, s\,(s - s'') + (s' - e)\,(2s' - e)\,] = 0 \qquad (5.2.44d)$$

from which it is apparent that at dwells, where all derivatives of $s(\psi)$ vanish, $H(\psi)$ vanishes as well, and the curvature attains stationary values.

The problem has then been reduced to finding the zeros of $H(\psi)$ as given by eq. (5.2.44d). These zeros can be found, again, either graphically with the aid of a graphics monitor and a *mouse*, as already explained above, or numerically.

Once the maximum absolute value of the curvature is obtained, the minimum radius of curvature is determined, and a corresponding value of a is computed as a percentage of the minimum radius of curvature, so as to avoid high contact stresses. Then, applying eq. (5.2.37), the cam profile for the roller-follower is generated. This is then used either to plot the profile or to generate the NC program needed to cut the cam profile, as outlined in Chapter 9.

5.2.7 Example 5.2.3: Cam Profile For a Translating Roller-Follower

From the cam profile generated in Example 5.2.1 for cycloidal motion, obtain the corresponding cam profile for a roller-follower. The roller radius should not exceed 50% of the minimum radius of curvature of the cam profile.

Solution:

Since we have already found the optimum values of c and e, for a knife-edge cam profile, we can plot function $H(\psi)$, given by eq. (5.2.44d), and shown in Fig. 5.2.12. From this plot, we obtain the values of ψ at which the radius of curvature attains stationary values, namely,

$$\psi = 59.24^\circ, \quad 92.17^\circ, \quad 274.73^\circ, \quad 307.94^\circ, \quad 328.48$$

Upon evaluation of κ_k at the foregoing values of ψ, we obtain, correspondingly,

$$\kappa_k = -0.0274, \quad 0.0174, \quad 0.0205, \quad 0.0206, \quad 0.0128$$

and hence, the global maximum of interest is $\kappa_M = 0.0274\text{mm}^{-1}$, which occurs at $\psi = 59.24^\circ$ at a convex point, i.e., at a point of negative curvature. From this value, the minimum radius of curvature was determined as $r_m = 36.49$mm. Thus, the roller radius was selected as $0.5 r_m$, which yields the follower radius as $a = 18.25$mm. The cam profile is shown in Fig. 5.2.13.

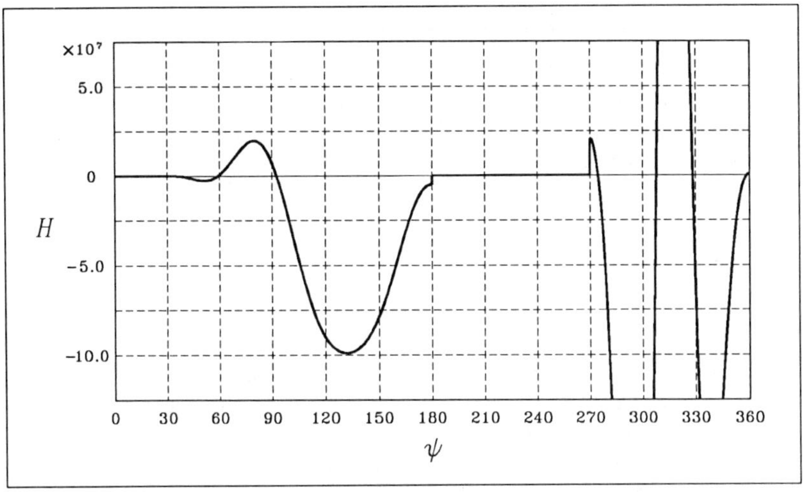

Fig. 5.2.12. Plot of $H(\psi)$ for Example 5.2.3.

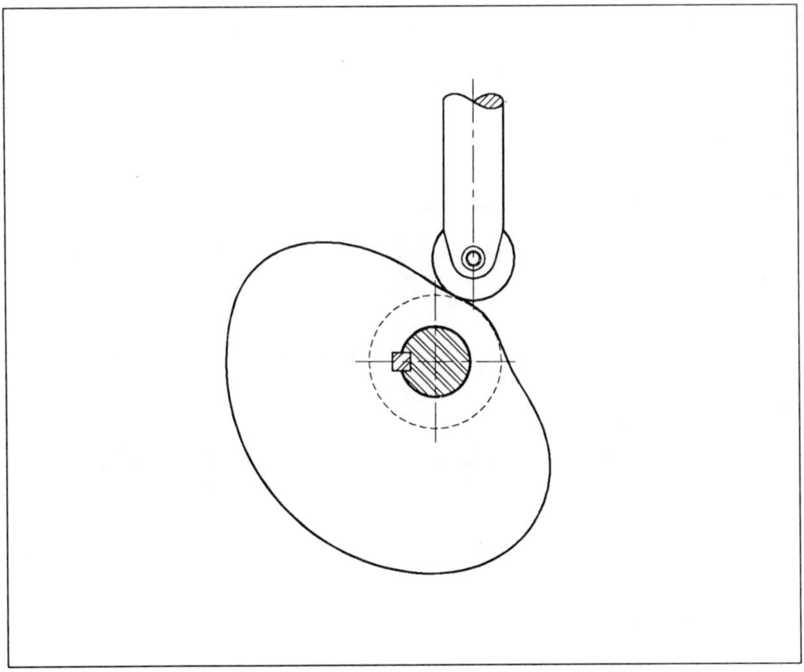

Fig. 5.2.13. Cam profile for a translating roller-follower.

5.3 Oscillating Roller-Followers

In this case, as in Section 5.2, it is assumed that the follower-displacement program is given as the sum of a constant β, as yet to be determined, and a positive-semidefinite function $\sigma(\psi)$, whose minimum value is 0, its maximum value being $\Delta\phi$, the amplitude of the follower oscillation, i.e.,

$$\phi(\psi) = \beta + \sigma(\psi) \tag{5.3.1a}$$

and hence,

$$\phi'(\psi) = \sigma'(\psi), \qquad \phi''(\psi) = \sigma''(\psi) \tag{5.3.1b}$$

The layout of the mechanism to be synthesized is shown in Fig. 5.3.1. In that figure, the pitch curve, appearing thinner, is the trajectory traced by the center of the roller on the cam disk. Alternatively, if the cam plate is machined with a cutter having a radius identical to that of its roller, then the aforementioned curve is the trajectory of the cutter center. From the same figure, segment $\overline{OP}(\equiv \ell)$ is attached to the fixed frame, whereas line OC is fixed to the cam disk. Moreover, segment $\overline{PR}(\equiv e)$ is the length of the follower arm, a parameter to be determined.

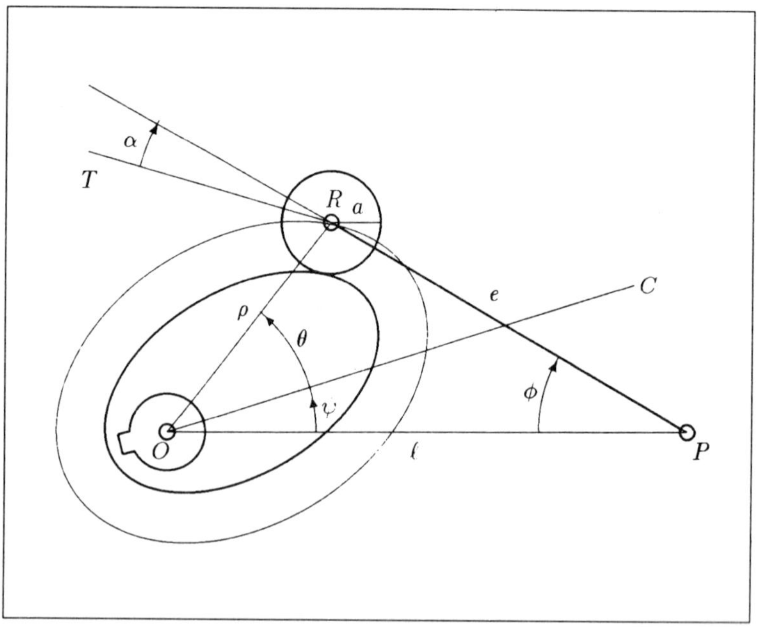

Fig. 5.3.1. Cam mechanism with oscillating roller-follower.

Let ψ and ϕ be the variables denoting the angular displacements of the cam and the follower, respectively, ρ and θ being the polar coordinates describing the cam profile. From the geometry of Fig. 5.3.2, the relations given below are readily derived, namely,

$$\rho \cos(\psi + \theta) = \ell - e \cos\phi \tag{5.3.2a}$$

$$\rho \sin(\psi + \theta) = e \sin\phi \tag{5.3.2b}$$

Combining eqs. (5.3.2*a* & *b*), the polar coordinates describing the cam profile are then obtained in parametric form as

$$\rho(\psi) = \sqrt{\ell + e^2 - 2\, e\, \ell \cos\phi} \tag{5.3.3a}$$

$$\theta(\psi) = \tan^{-1}\left(\frac{e \sin\phi}{\ell - e\cos\phi}\right) - \psi \tag{5.3.3b}$$

Alternatively, the Cartesian coordinates x and y are computed as explained below. From eqs. (5.3.2*a* & *b*),

$$\rho\cos\theta\cos\psi - \rho\sin\theta\sin\psi = \ell - e\,\cos\phi \tag{5.3.4a}$$

$$\rho\cos\theta\sin\psi + \rho\sin\theta\cos\psi = e\,\sin\phi \tag{5.3.4b}$$

Since $\rho\cos\theta = x$ and $\rho\sin\theta = y$ are the Cartesian coordinates of the cam profile in a frame fixed to the cam plate, eqs. (5.3.4*a* & *b*) can be rewritten as

$$\mathbf{Q}(\psi)\,\mathbf{p}(\psi) = \mathbf{r}(\psi) \tag{5.3.5a}$$

where $\mathbf{Q}(\psi)$, $\mathbf{p}(\psi)$, and $\mathbf{r}(\psi)$ are given by

$$\mathbf{Q}(\psi) = \begin{bmatrix} \cos\psi & -\sin\psi \\ \sin\psi & \cos\psi \end{bmatrix}, \quad \mathbf{p}(\psi) = \begin{bmatrix} x(\psi) \\ y(\psi) \end{bmatrix}, \quad \mathbf{r}(\psi) = \begin{bmatrix} \ell - e\cos\phi \\ e\sin\phi \end{bmatrix} \tag{5.3.5b}$$

$\mathbf{Q}(\psi)$ being an orthogonal matrix, as already discussed, while $\mathbf{r}(\psi)$ is the position vector of the contact point in a coordinate system fixed to the mechanism frame, with axes along the horizontal and vertical directions of Fig. 5.3.1 and origin at point O of the same figure. We will thus define

$$\xi \equiv \ell - e\cos\phi, \qquad \eta \equiv e\sin\phi \tag{5.3.5c}$$

and hence,

$$\mathbf{r}(\psi) = \begin{bmatrix} \xi \\ \\ \eta \end{bmatrix} \tag{5.3.5d}$$

From eq. (5.3.5a), $\mathbf{p}(\psi)$ is obtained as $\mathbf{Q}^T(\psi)\,\mathbf{r}(\psi)$, which yields the Cartesian coordinates as

$$x(\psi) = \xi \cos\psi + \eta \sin\psi = \ell \cos\psi - e\,\cos(\psi + \phi) \tag{5.3.6a}$$

$$y(\psi) = \eta \cos\psi - \xi \sin\psi = -\ell \sin\psi + e\,\sin(\psi + \phi) \tag{5.3.6b}$$

5.3.1 Cam-Size Optimization Under a Pressure Angle Constraint

In order to obtain a good torque transmission, as discussed in Subsection 1.4.1, the pressure angle α, shown in Fig. 5.3.2 is to be bounded, namely,

$$-\alpha_M \le \alpha \le \alpha_M$$

where α_M is the maximum allowable value of α. However, determining the value of the cam angle of rotation ψ at which α attains its extrema is not straightforward, for, as in the case of translating knife-edge and roller followers, this variable is not readily available. Hence, the extrema of a smooth-enough function of α, namely, $\tan\alpha$, are sought; that is, the values ψ^* of ψ, at which $\tan\alpha$ attains its extrema, are to be determined. To this end, an expression for $\tan\alpha$ is first derived.

From the geometry of Fig. 5.3.1, one has

$$\gamma + \alpha = \pi - (\psi + \theta + \phi) \tag{5.3.7a}$$

Thus,

$$\tan\alpha = \tan(\psi + \theta + \phi + \gamma) \tag{5.3.7b}$$

Resorting to the trigonometric relation for the tangent of a sum, applied to the case at hand, we derive

$$\tan(\psi + \theta + \phi + \gamma) = \frac{\tan(\psi + \theta + \phi) + \tan\gamma}{1 - \tan(\psi + \theta + \phi)\,\tan\gamma} \tag{5.3.8a}$$

where

$$\tan(\psi + \theta + \phi) = \frac{\tan(\psi + \theta) + \tan\phi}{1 - \tan(\psi + \theta)\tan\phi} \tag{5.3.8b}$$

From eq. (3.2.4b), $\tan\gamma$ is computed as

$$\tan\gamma = \frac{\rho(\theta)}{\rho'(\theta)} \tag{5.3.8c}$$

Furthermore, upon substitution of eqs. (5.3.8a–c) into eq. (5.3.7b), we obtain

$$\tan\alpha = \frac{\rho'(\theta)\,[\tan(\psi+\theta)+\tan\phi] + \rho\,[1-\tan(\psi+\theta)\tan\phi]}{\rho'(\theta)\,[1-\tan(\psi+\theta)\tan\phi] - \rho\,[\tan(\psi+\theta)+\tan\phi]} \tag{5.3.9}$$

On the other hand, combining eqs. (5.3.2a & b), one obtains

$$\tan(\psi+\theta) = \frac{e\sin\phi}{\ell - e\cos\phi} = \frac{\xi}{\eta} \tag{5.3.10}$$

Moreover, substitution of eq. (5.3.10) into eq. (5.3.9) yields

$$\tan\alpha = \frac{\rho'(\theta)\,[e\sin\phi + (\ell - e\cos\phi)\tan\phi] + \rho\,[\ell - e\cos\phi - e\sin\phi\tan\phi]}{\rho'(\theta)\,[\ell - e\cos\phi - e\sin\phi\tan\phi] - \rho\,[e\sin\phi + (\ell - e\cos\phi)\tan\phi]} \tag{5.3.11}$$

or, in terms of $\cos\phi$ and $\sin\phi$,

$$\tan\alpha = \frac{\ell\,\rho'(\theta)\sin\phi + \rho\,(\ell\cos\phi - e)}{\rho'(\theta)\,(\ell\cos\phi - e) - \ell\,\rho\sin\phi} \tag{5.3.12}$$

Now, $\rho'(\theta)$ is computed as

$$\rho'(\theta) = \frac{\rho'(\psi)}{\theta'(\psi)} \tag{5.3.13}$$

where $\rho'(\psi)$ and $\theta'(\psi)$ are derived below.

Differentiation of both sides of eqs. (5.3.2a & b) with respect to ψ yields

$$\rho'(\psi)\cos(\psi+\theta) - \rho\,[1+\theta'(\psi)]\sin(\psi+\theta) = e\phi'(\psi)\sin\phi \tag{5.3.14a}$$

$$\rho'(\psi)\sin(\psi+\theta) + \rho\,[1+\theta'(\psi)]\cos(\psi+\theta) = e\phi'(\psi)\cos\phi \tag{5.3.14b}$$

and, from eqs. (5.3.2a & b), one has

$$\cos(\psi+\theta) = \frac{\xi}{\rho}, \qquad \sin(\psi+\theta) = \frac{\eta}{\rho} \tag{5.3.14c}$$

with ξ and η defined in eq. (5.3.5c). Upon substitution of eqs. (5.3.14c) into eqs. (5.3.14a & b) and reduction of the ensuing relations, eqs. (5.3.14a & b) can be rewritten in the form

$$\mathbf{A}\,\mathbf{x} = \mathbf{b} \tag{5.3.15a}$$

where

$$\mathbf{A} = \frac{1}{\rho}\begin{bmatrix} \xi & -\rho\eta \\ \eta & \rho\xi \end{bmatrix}, \tag{5.3.15b}$$

$$\mathbf{x} = \begin{bmatrix} \rho' \\ \theta' \end{bmatrix}, \qquad \mathbf{b} = \begin{bmatrix} (1+\phi')\,\eta \\ \ell\,\phi' - (1+\phi')\,\xi \end{bmatrix} \tag{5.3.15c}$$

the argument ψ having been dropped for compactness and all derivatives are understood to be taken with respect to ψ.

From eq. (5.3.15a), $\mathbf{x}$ is obtained as $\mathbf{A}^{-1}\mathbf{b}$, namely,

$$\begin{bmatrix} \rho' \\ \theta' \end{bmatrix} = \frac{1}{\rho^2}\begin{bmatrix} \rho\xi & \rho\eta \\ -\eta & \xi \end{bmatrix}\begin{bmatrix} (1+\phi')\,\eta \\ \ell\,\phi' - (1+\phi')\,\xi \end{bmatrix} \tag{5.3.15d}$$

Thus,

$$\rho'(\psi) = \frac{\eta\,\ell\,\phi'(\psi)}{\rho}, \tag{5.3.16a}$$

$$\theta'(\psi) = \frac{\ell\,\xi\,\phi' - \rho^2\,(1+\phi')}{\rho^2} \tag{5.3.16b}$$

Substitution of eqs. (5.3.16a & b) into eq. (5.3.13) yields

$$\rho'(\theta) = \frac{\rho\,\eta\,\ell\,\phi'(\psi)}{\ell\,\xi\,\phi' - \rho^2\,(1+\phi')} \tag{5.3.17}$$

which, upon substitution into eq. (5.3.12) and reduction of the ensuing equations, yields

$$\tan\alpha = \frac{e\,[\,1-\phi'(\psi)\,] - \ell\cos\phi}{\ell\sin\phi} \tag{5.3.18}$$

thereby deriving the relation sought.

Recalling eqs. (5.3.1a & b), eq. (5.3.18) can be rewritten as

$$\tan\alpha = \frac{u\,[\,1-\sigma'(\psi)\,] - \cos(\beta+\sigma)}{\sin(\beta+\sigma)} \tag{5.3.19}$$

where $u \equiv e/\ell$. Moreover, the follower at its lowest position is shown in Fig. 5.3.2. In this position, the pressure angle will be assumed to attain an absolute value α_M. From the geometry of that figure, one has

$$\sin\beta = v\cos\alpha_M, \qquad \cos\beta = u + v\sin\alpha_M \tag{5.3.20}$$

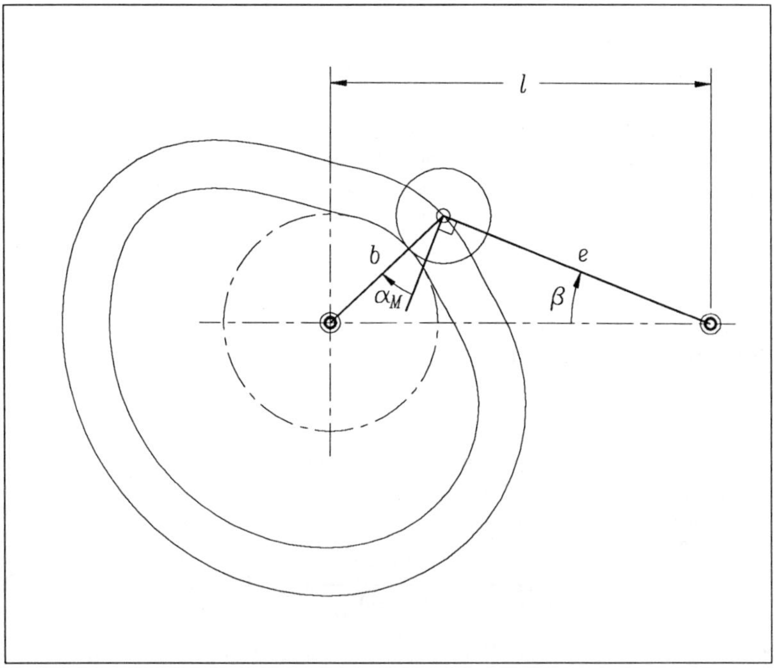

Fig. 5.3.2. Follower at its lowest position.

where $v \equiv b/\ell$. Substitution of eq. (5.3.20) into eq. (5.3.19) leads to

$$\tan\alpha = \frac{1+\sigma'(\psi) - \cos\sigma + w\,\sin(\sigma - \alpha_M)}{\sin\sigma + w\,\cos(\sigma - \alpha_M)} \equiv \frac{N(\psi)}{D(\psi)} \tag{5.3.21}$$

with w defined as

$$w \equiv \frac{v}{u} = \frac{b}{e}$$

Now, the extrema of $\tan\alpha$ are found by zeroing its first derivative with respect to ψ, namely,

$$\frac{d\tan\alpha}{d\psi} = \frac{1}{D}\,[\,N'(\psi) - \tan\alpha\,D'(\psi)\,] = 0 \tag{5.3.22a}$$

which is called the *normality condition* for $\tan\alpha$. In this equation,

$$N'(\psi) \equiv \sigma'' + \sigma'\,[\sin\sigma + w\,\cos(\sigma - \alpha_M)\,], \tag{5.3.22b}$$

$$D'(\psi) \equiv \sigma'\,[\cos\sigma - w\cos(\sigma - \alpha_M)\,] \tag{5.3.22c}$$

Thus, from the normality condition (5.3.22a), one has

$$\sigma'' + \sigma'\{[\sin\sigma + w\cos(\sigma - \alpha_M)] - \tan\alpha_M[\cos\sigma - w\cos(\sigma - \alpha_M)]\} = 0 \quad (5.3.22d)$$

where the argument ψ has been dropped for brevity.

Let ψ^*, as yet undetermined, be the value at which the pressure angle attains its maximum absolute value during the rise phase. Then, from eq. (5.3.21), one obtains

$$w = \frac{(1+\sigma_1')\cos\alpha_M - \cos(\alpha_M - \sigma_1)}{\sin(2\alpha_M - \sigma_1)} \quad (5.3.23)$$

where $\sigma_1 \equiv \sigma(\psi^*)$ and $\sigma_1' \equiv \sigma'(\psi^*)$.

Substitution of eq. (5.3.23) into the normality condition (5.3.22d) yields

$$G(\psi) \equiv \sin(2\alpha_M - \sigma)\sigma'' + [(1+\sigma')\cos(2\alpha_M - \sigma) - 1]\sigma' = 0 \quad (5.3.24)$$

Thus, the optimization problem has been reduced to finding the value ψ^* that satisfies eq. (5.3.24). Once this value has been found, w is computed using eq. (5.3.23), while u and v are computed as

$$u = \frac{1}{\sqrt{1 + w^2 + 2w\sin\alpha_M}}, \qquad v = uw \quad (5.3.25)$$

and finally, for a given value of ℓ, the optimum radius of the base circle, b_0, is computed. An example implementing the above methodology is given below.

5.3.2 Example 5.3.1: Pitch-Curve of an Oscillating Roller-Follower

Synthesize the pitch curve of a cam-follower mechanism for an oscillating roller-follower that will produce the follower-displacement program appearing in Table 5.3.1. The motion is prescribed as cycloidal, and the pressure angle is to attain a maximum absolute value of 30°, the amplitude of the follower oscillations being prescribed as 45°.

Table 5.3.1: Prescribed angular-displacement program.

Phase	Angle of rotation(ψ)	Displacement (ϕ)
D_1(dwell)	36°	0°
R_1(rise)	144°	+45°
D_2(dwell)	90°	0°
R_2(return)	90°	−45°

Solution:

The cam profile is now designed following the methodology presented above. First, the value of ψ^* is obtain by plotting function $G(\psi)$, as given by eq. (5.3.24). This plot is displayed in Fig. 5.3.3, from which we obtained $\psi^* = 129.4°$. Then, from eq. (5.3.23), w is found with this value of ψ^* as $w = 0.7115$, and hence,

$$u = 0.6715, \quad v = 0.4778$$

thereby obtaining

$$\beta = \tan^{-1}\left(\frac{v \cos\alpha_M}{u + v \sin\alpha_M}\right) =$$

$$= \tan^{-1}\left(\frac{w \cos\alpha_M}{1 + w \sin\alpha_M}\right) = 0.4266 \text{ rad} = 24.44°$$

Thus, for a unit value of ℓ, $e = u = 0.6715$, and the base radius $b_0 = v = 0.4778$. The Cartesian coordinates are then computed using eqs. (5.3.6*a* & *b*), the synthesized pitch curve being shown in Fig. 5.3.4.

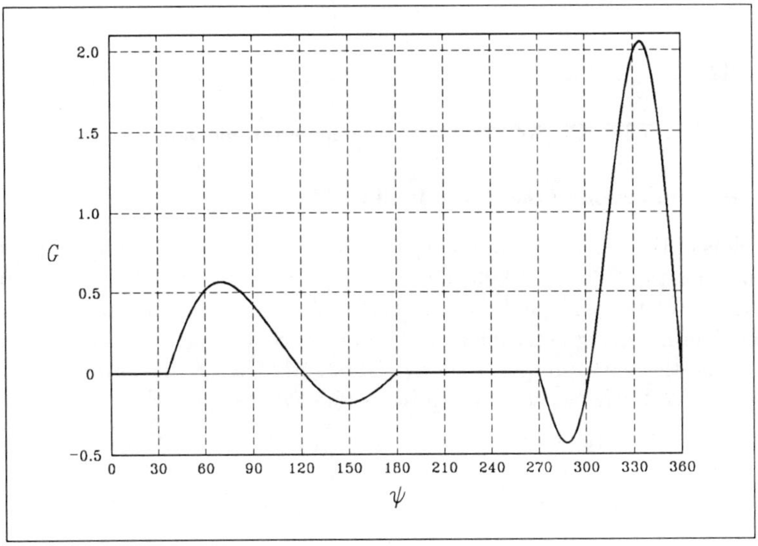

Fig. 5.3.3. Function $G(\psi)$ as given by eq. (5.3.24).

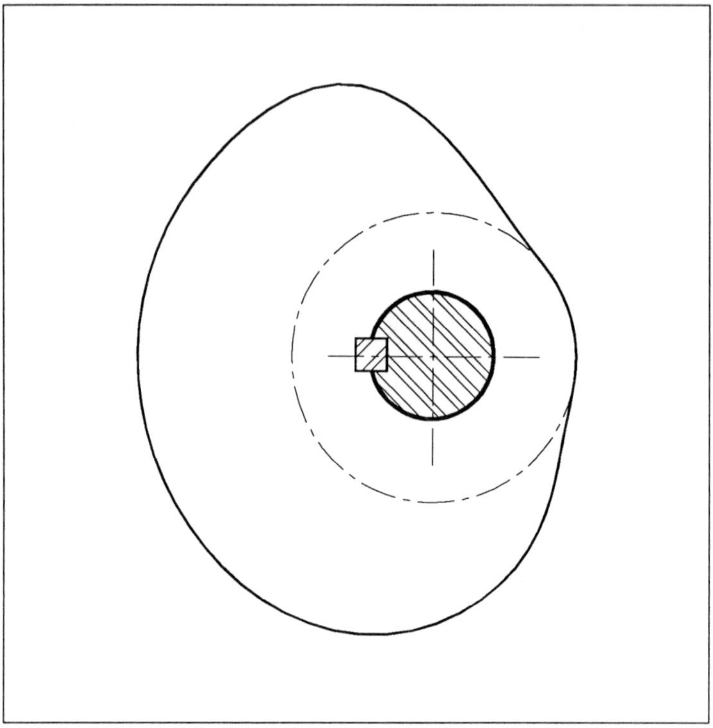

Fig. 5.3.4. Plot of the pitch-curve for Example 5.3.1, for $\ell = 1$.

5.3.3 Determination of the Roller Radius

The radius of the roller for a unit length ℓ, a/ℓ can be computed from the pitch curve for a unit length ℓ, by prescribing a/ℓ to be a given fraction ν of the minimum radius of curvature of the said pitch curve. This is done in order to avoid undercutting, the said fraction being determined, in turn, from contact-stress considerations. The formula for the curvature κ_k is derived presently, using the formulas derived in Subsection 3.2.2, where $\mathbf{r}(\psi)$ is given in eq. (5.3.5*b*), its derivatives being

$$\mathbf{r}'(\psi) = \begin{bmatrix} e\,\phi'(\psi)\sin\phi \\ e\,\phi'(\psi)\cos\phi \end{bmatrix}, \qquad \mathbf{r}''(\psi) = \begin{bmatrix} e\,(\phi'^2\cos\phi + \phi''\sin\phi) \\ e\,(\phi''\cos\phi - \phi'^2\sin\phi) \end{bmatrix} \tag{5.3.26}$$

Thus,

$$\mathbf{r}'(\psi) - \mathbf{E}\,\mathbf{r}(\psi) = \begin{bmatrix} e\,(1+\phi')\sin\phi \\ e\,(1+\phi')\cos\phi - \ell \end{bmatrix} \tag{5.3.27a}$$

while

$$\mathbf{r}'' - 2\,\mathbf{E}\,\mathbf{r}' - \mathbf{r} = \begin{bmatrix} e\,\left[(1+\phi')^2\cos\phi + \phi''\sin\phi\right] - \ell \\ e\,\left[-(1+\phi')^2\sin\phi + \phi''\cos\phi\right] \end{bmatrix} \tag{5.3.27b}$$

Therefore,

$$\mathbf{E}\,(\mathbf{r}'' - 2\,\mathbf{E}\,\mathbf{r}' - \mathbf{r}) = \begin{bmatrix} -e\,\left[-(1+\phi')^2\sin\phi + \phi''\cos\phi\right] \\ e\,\left[(1+\phi')^2\cos\phi + \phi''\sin\phi\right] - \ell \end{bmatrix} \tag{5.3.27c}$$

Thus, after simplifications,

$$\mathbf{p}'^T\,\mathbf{E}\,\mathbf{p}'' = (\mathbf{r}' - \mathbf{E}\,\mathbf{r})^T\mathbf{E}\,(\mathbf{r}'' - 2\,\mathbf{E}\,\mathbf{r}' - \mathbf{r}) =$$

$$= e^2\,(1+\phi')^3 - e\,\ell\,[\,(1+\phi')\,(2+\phi')\cos\phi + \phi''\sin\phi\,] + \ell^2 \tag{5.3.27d}$$

and

$$\|\,\mathbf{p}'\,\|^2 = \|\mathbf{Q}^T(\psi)\,[\mathbf{r}'(\psi) - \mathbf{E}\,\mathbf{r}(\psi)\,]\,\|^2 = \|\,\mathbf{r}'(\psi) - \mathbf{E}\,\mathbf{r}(\psi)\,\|^2 =$$

$$= e^2(1+\phi')^2 - 2\,e\,\ell\,(1+\phi')\cos\phi + \ell^2 \tag{5.3.27e}$$

Upon substitution of eqs. (5.3.27*d* & *e*) into eq. (3.2.16), the desired expression for κ_k reduces to

$$\kappa_k = \frac{e^2\,(1+\phi')^3 - e\,\ell\,[\,(1+\phi')\,(2+\phi')\cos\phi + \phi''\sin\phi\,] + \ell^2}{[\,e^2\,(1+\phi')^2 - 2\,e\,\ell\,(1+\phi')\cos\phi + \ell^2\,]^{3/2}} \equiv \frac{N_k}{D_k} \tag{5.3.28}$$

From eq. (5.3.28) one can compute the maximum absolute value of the curvature, κ_M. This can be done either numerically or graphically, as in previous cases. Fig. 5.3.5 shows a plot of the curvature, κ_k, for the conditions given for Example 5.3.1.

Once the value of the maximum curvature has been obtained, one can produce the actual cam profile for the roller-follower as explained below. Let r_m be the minimum value of r/ℓ, $\mathbf{r}_k$, $\mathbf{r}_r$, and $\mathbf{e}_n$ denoting the position vectors of corresponding points on the pitch curve and on the cam profile, as well as the unit normal vector of both curves at corresponding points, respectively. The cam profile is thus synthesized, as explained in Section 5.2, from the relation

$$\mathbf{r}_r = \mathbf{r}_k + a\,\mathbf{e}_n$$

where a is the radius of the roller. The cam profile obtained from the above equation corresponds to a certain value of the parameter a per unit length ℓ. The latter can be chosen, in turn, from considerations of space availability, together with the maximum allowable value of the contact stress. Once this parameter is determined, the actual cam profile is obtained by scaling the foregoing normalized parameters.

The above formulation can be readily implemented in a code. Its input would be the displacement program, given in the form of percentages, and the maximum pressure angle α_M. The minimum radius of curvature is computed after finding the maximum curvature. The final cam profile is obtained as explained in Subsection 5.2.3.

5.3.4 Design Procedure

The overall design procedure is summarized below in algorithmic form:

1. Find ψ^* from the plot of function $G(\psi)$ as given by eq. (5.3.24); then,
2. compute w with the value obtained for ψ^*; next,
3. compute u, v, and the base radius b_0 for a given ℓ using relations (5.3.25) and the definitions of the above variables. Further,
4. generate the displacement program $\phi(\psi)$ with the above variables. Now,
5. compute the Cartesian coordinates of the cam profile using eqs. (5.3.6a & b). Next,
6. determine the maximum radius of curvature by using eq. (5.3.27) or, alternatively, by plotting the same equation as explained in previous sections. Finally,
7. determine the cam profile for the roller-follower using eq. (5.3.28).

5.3.5 Example 5.3.2: Cam Profile for an Oscillating Roller-Follower

Determine the cam profile for the conditions given in Example 5.3.1 if the maximum radius of the roller is to be 50% of the minimum radius of curvature.

Solution:

The data obtained in Example 5.3.1, namely, $e/\ell = 0.6715$, $b_0/\ell = 0.4778$, and $\beta = 0.4266$ rad, along with the values for displacement, velocity, and acceleration are plugged into eq. (5.3.27). These produce the curvature distribution displayed in Fig. 5.3.5. This plot shows four local extrema during the rise and return phases, which are

$$\kappa_1 = -0.348, \quad \kappa_2 = 1.68, \quad \kappa_3 = 2.72, \quad \kappa_4 = -0.577$$

From these values the minimum radius of curvature is determined, which is $r_m = 1/2.72 = 0.37$. Therefore, the follower radius is chosen as

$$\frac{a}{\ell} = 0.50 \times 0.37 = 0.185$$

With the above information and a given value of ℓ, the Cartesian coordinates for the actual cam profile are computed using eq. (5.3.28). This is shown in Fig. 5.3.6, for a unit value of ℓ.

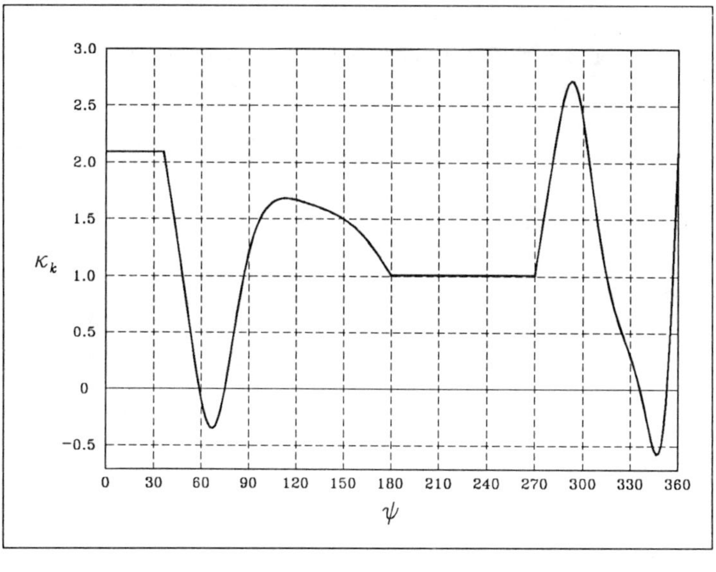

Fig. 5.3.5. Plot of the curvature for Example 5.3.1.

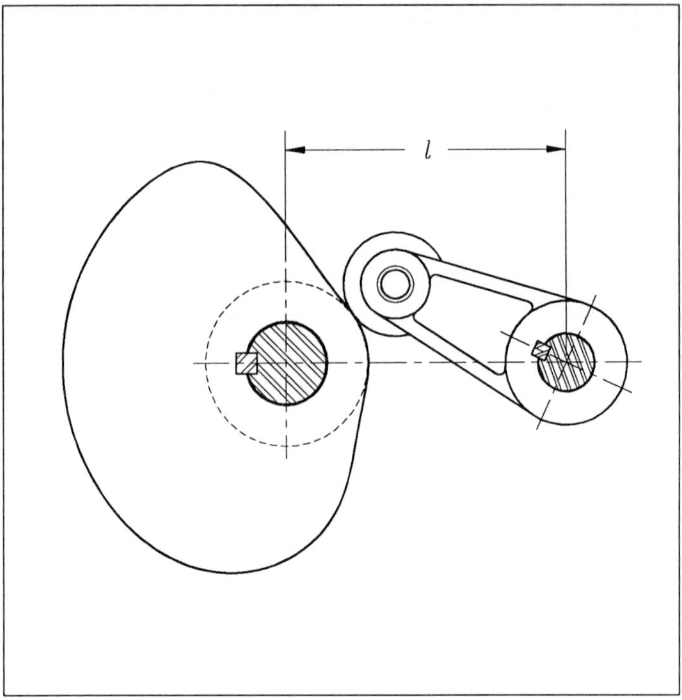

Fig. 5.3.6. Cam profile for Example 5.3.2, for $\ell = 1$.

Chapter 6

AN INTRODUCTION TO SPATIAL CAMS

6.1 Introduction

As outlined in Chapter 1, spatial cams are by far the most varied type, and, for this reason, their classification is very difficult. We will focus on a very specific type of spatial cam mechanism in this chapter, while trying to maintain the spirit of Chapters 4 and 5. We will devote this chapter to *cylindrical cams* with a translating follower, a solid model of which is shown in Fig. 6.1.1.

Now, if we tried to parallel Chapters 4 and 5 exactly, we would have to study flat-face and roller-followers in the realm of cylindrical cams. However, it turns out that the cylindrical counterpart of the disk cam with a flat-face follower is physically impractical, although we would not go as far as saying that it is *physically impossible*. On the other hand, cylindrical cams with *oscillating* roller followers pose technical problems that lie beyond the scope of this book and hence, this type of cylindrical cam will be omitted. Thus, we will limit this chapter to cylindrical cams with translating roller-followers. We use a cylindrical cam with a translating roller-follower whenever we need to transmit motion from a turning shaft to an element translating in a direction parallel to the shaft axis. Motion transmission in this form is very efficient if the roller is designed judiciously. In fact, by properly designing the roller, we can produce pure-rolling motion among the cam, the roller, and the follower. However, the follower moves with pure translation with respect to the machine frame, and hence friction forces between these two elements become important. Moreover, reaction forces at the bearings can be significantly high and hence may impinge on the life span of the machine. We can properly keep bearing forces of a static nature within bounds by designing the cam track accordingly, in order to keep the pressure angle between suitable bounds.

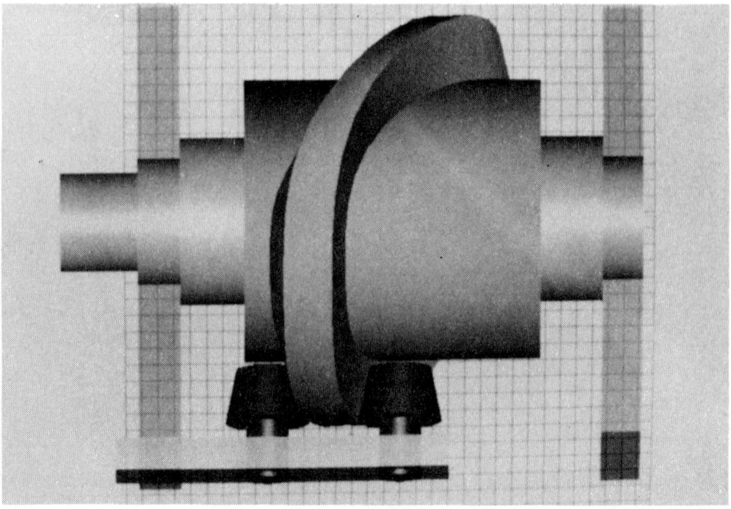

Fig. 6.1.1. Solid model of a cylindrical cam with a translating roller-follower (produced by Max A. González-Palacios).

The synthesis of the type of cam discussed here is quite straightforward and has been studied in the specialized literature (Rothbahrt, 1956). Its optimization in the spirit of Chapters 4 and 5 is quite simple as well, although it has not been reported so far, to the best of our knowledge.

6.2 Cylindrical Cams with a Translating Roller-Follower

A typical layout of a cylindrical cam of this type is shown in Fig. 6.2.1. Just as for the planar counterpart of this type of cam, studied in Section 5.2, we distinguish four rigid elements, namely, the frame, the cam cylinder, the roller, and the follower. The counterpart of the cam profile in the case of disk cams is the cam *track* in the case at hand. Because of practical considerations, to be discussed below, a cam cylinder with a *protruded track* offers advantages over that with a *grooved track*. Below we discuss the geometrical shape of the track and the roller for pure-rolling motion.

6.2.1 Geometry of Track and Roller

We discuss here the rationale behind the shaping of the track and the roller. To this end, we resort to the spatial kinematics of rigid bodies; in particular, we invoke the Aronhold-Kennedy Theorem in three dimensions, first proposed by Beggs (1959).

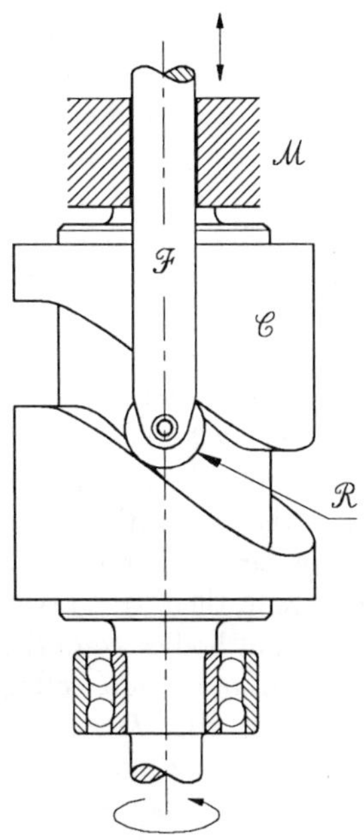

Fig. 6.2.1. Elements of a cylindrical cam.

Central to the Aronhold-Kennedy Theorem is the concept of *instant axis of relative motion* or, simply, *instant axis*. The instant axis of two rigid bodies under relative motion is the locus of all points instantaneously common to the two bodies whose relative velocity has a minimum magnitude. Thus, let A denote a point of body $\mathcal{A}$ and B a point of body $\mathcal{B}$, their position vectors being denoted by $\mathbf{a}$ and $\mathbf{b}$, respectively. We assume further that A and B coincide at a certain instant, i.e., that $\mathbf{a} = \mathbf{b}$ instantaneously. However, since these two points belong to different rigid bodies, their velocities do not necessarily coincide. Thus, the relative velocity of these points, $\mathbf{v}_{AB}$, defined below, is generally not zero.

$$\mathbf{v}_{AB} \equiv \dot{\mathbf{a}} - \dot{\mathbf{b}} \tag{6.2.1}$$

Next, we look for all those points A and B of these two bodies whose relative velocity has a minimum magnitude. It turns out (Hunt, 1978; Angeles, 1988) that the locus of all such points is a line that is called the instant screw axis, or simply the instant axis, of body $\mathcal{A}$ with respect to body $\mathcal{B}$. Pure rolling occurs when the velocity of minimum magnitude vanishes.

Furthermore, according to the Aronhold-Kennedy Theorem, pure-rolling arises among all three possible pairs of a set of three given rigid bodies in relative motion if, and only if, the instant axes of relative motion between any two of those pairs intersect. In other words, if the three bodies are denoted by $\mathcal{A}$, $\mathcal{B}$, and $\mathcal{C}$ while the instant axes of every pair of these bodies are denoted by I_{AB}, I_{BC}, and I_{CA}, then, under the assumption that $\mathcal{A}$ and $\mathcal{B}$, $\mathcal{B}$ and $\mathcal{C}$, as well as $\mathcal{C}$ and $\mathcal{A}$ move with respect to each other in pure-rolling motion, I_{AB}, I_{BC}, and I_{CA} are concurrent at one single point and coplanar.

In this study, the three bodies under relative motion are the cam cylinder, labeled $\mathcal{C}$; the roller, labeled $\mathcal{R}$; and the follower, labeled $\mathcal{F}$. A schematic layout of these bodies is illustrated in Fig. 6.2.2. In this figure, the instant axis of $\mathcal{R}$ with respect to $\mathcal{F}$, labeled I_{RF}, is simply the axis of rotation of the roller with respect to the follower, whereas that of the cam with respect to the follower passes through the point of contact Q between cam and roller and is denoted by I_{CR}. By virtue of the Aronhold-Kennedy Theorem, then, these two axes must intersect. Moreover, if we look now at the triad of rigid bodies comprising the machine frame, $\mathcal{M}$, the cam, and the roller, these are also under pure rolling, and hence, the two foregoing intersecting axes must be concurrent with I_{MC}, which is merely the axis of rotation of the cylinder cam with respect to the machine frame. Thus, I_{FC} is concurrent with these axes as well.

The layout depicted in Fig. 6.2.2 is general and hence, valid for any possible configuration of the mechanism. Hence, while the mechanism is in operation, the point of concurrence of all four instant axes discussed so far moves along I_{MC}, which is fixed on the machine frame. In the course of this operation, I_{CR} generates a circular conical surface with axis I_{RF} on the roller, while I_{CR} traces a more complex ruled surface on the cam cylinder. This surface is called the *cylindroid*—see (Hunt, 1978) for a detailed discussion and some historical notes on this concept. As a consequence, for pure rolling as described above, the roller should be shaped as a circular cone with angle β, while the track on the cylinder should be shaped as a cylindroid. A surface with the shape of a cylindroid is depicted in Fig. 6.2.3.

6.2.2 Cam Optimization Based on Minimum-Size Cylinder and Maximum Pressure Angle

From the above discussion, the roller is a circular cone and the track on the cam cylinder a cylindroid. However, these are unbounded surfaces and hence, we must proceed to define bounds for them. First and foremost, the cylindroid is in fact an unbounded surface whose shape is dictated by the displacement program of the follower. Moreover, whereas Fig. 6.2.4 depicts only the *pitch curve* of the cam, the cam cylinder must be given dimensions for manufacture, i.e., we must bound the originally unbounded cylindroid. This is done with an inner and an outer cylinder, as shown in Fig. 6.2.5.

Note that the pitch curve is inscribed on a circular cylinder, called the *pitch cylinder*. However, the element of the roller that is in contact with the track does not maintain a constant orientation with respect to I_{MC}. Let R and S be the two end points of the roller element in contact with the track, which are indicated on the track in Fig. 6.2.5. Moreover, the orientation of the plane of the three coplanar axes, I_{CR}, I_{RF}, and I_{FC}, changes with respect to I_{MC} when the mechanism is in operation. This means that the projection of segment RS onto a plane perpendicular to I_{MC} is not of constant length, and, as a consequence, the two bounding cylinders shown in Fig. 6.2.4 *are not circular*. Clearly, machining them to their shapes dictated by the motion of points R and S becomes more costly than if they were circular cylinders.

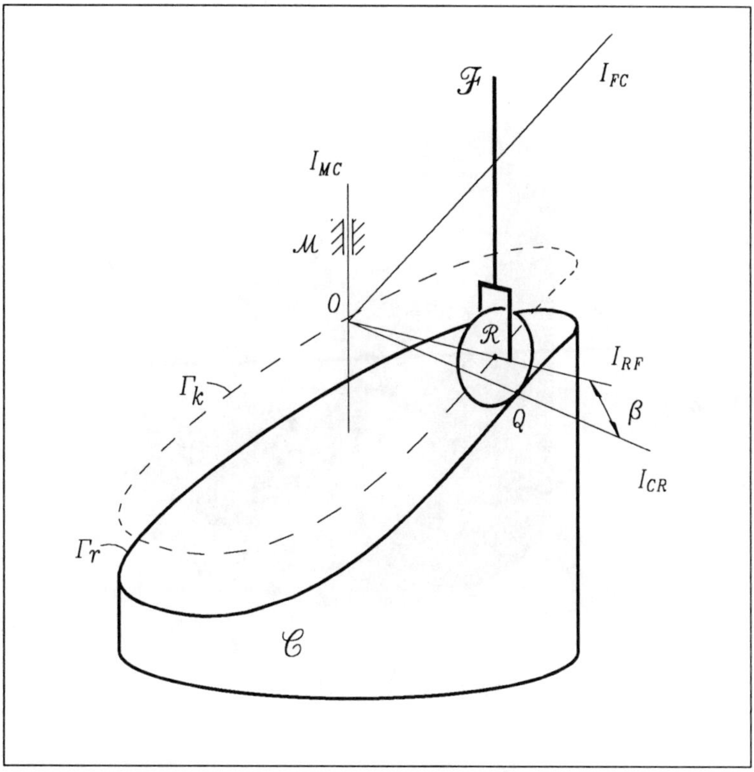

Fig. 6.2.2. Geometry of a cylindrical cam.

However, nothing prevents us from shaping the two bounding cylinders as circular cylindrical surfaces, if they are given the suitable radii. In order to use the full length of the contact segment RS, it is then convenient to dimension the hollow cylinder so that its wall thickness will be at least equal to the maximum projection of the length of segment RS onto a plane perpendicular to I_{MC}. This maximum projection is obviously given by $d \cos \beta$, where d is the length of segment RS.

The shape of the cylindroid is, of course, dictated by the prescribed motion of the follower. However, the *pitch cylinder*, depicted in Fig. 6.2.2, can be chosen arbitrarily through its sole geometric parameter, namely its radius a. Because of space limitations, economy of materials, and concerns about dynamical forces, we would like to dimension a as small as possible with respect to a certain characteristic length of the task at hand. Obviously, this length is the rise h. Now, if the ratio a/h is made too small, then the maximum absolute value of the pressure angle can attain inadmissibly large values, and hence an optimum ratio a/h must be chosen so as to keep the pressure angle within certain bounds. The optimum choice of this ratio is discussed below.

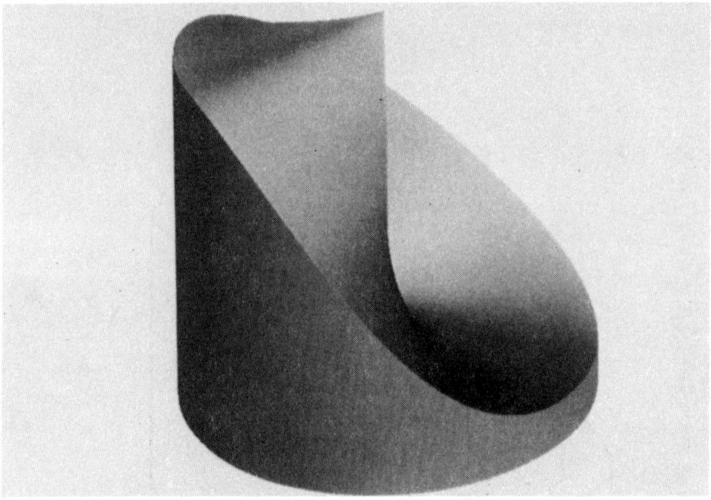

Fig. 6.2.3. The cylindroid.

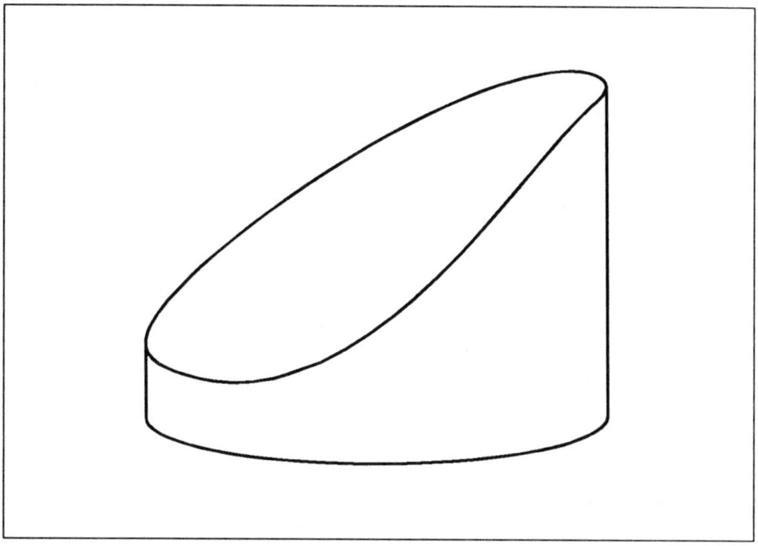

Fig. 6.2.4. Pitch curve of a cylindrical cam.

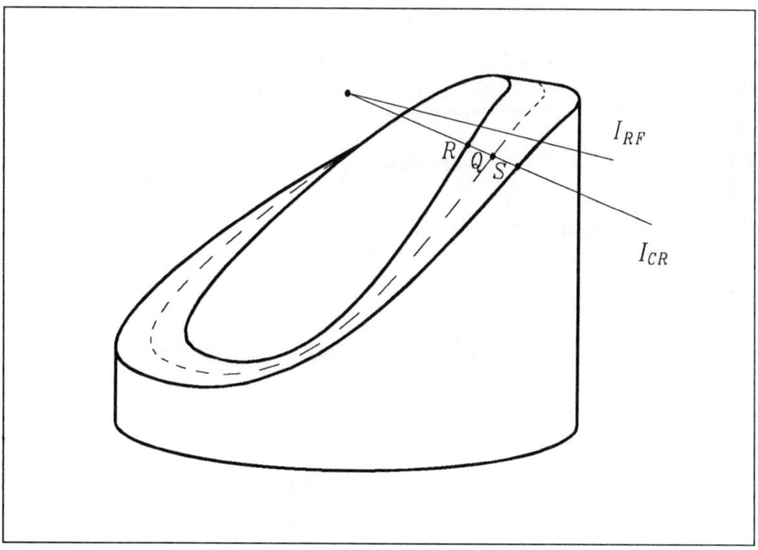

Fig. 6.2.5. Bounds of the inner and outer cylinders of a cylindrical cam.

First, the pitch curve Γ_k of the equivalent knife-edge follower is determined. This curve will help us determine the radius of the roller-follower based on bounds on the pressure angle. In fact, if Γ_r denotes the *pitch curve* for the roller-follower, then the pressure angle is the same for both pitch curves, Γ_k and Γ_r, as illustrated in Fig. 6.2.6. Note that Γ_k is merely the intersection of the pitch cylinder with a cylindroid that is swept by a line intersecting I_{MC} at right angles. Moreover, this line is no other than I_{RF}.

Next, the pitch curve Γ_k is developed, as shown in Fig. 6.2.7, in which ξ denotes the length of a circular arc of radius a and $\sigma(\psi)$ is a positive-semidefinite function of the angle of rotation of the cam cylinder, ψ, defining the displacement program of the follower. Note that the constant c added to σ in the case of a disk cam with a translating knife-edge follower becomes irrelevant in this case, and hence, the only design parameter to be determined is a. Note also that the follower motion takes place along a direction parallel to the σ axis in the σ vs. ξ plane. Hence, the pressure angle α turns out to be the angle between a line parallel to the σ axis and the normal to the σ-vs.-ξ curve, as shown in Fig. 6.2.7. This angle is thus simply the one that the tangent to the same curve makes with the ξ axis, i.e.,

$$\tan\alpha = \frac{d\sigma}{d\xi} \tag{6.2.2}$$

where ξ is given by

$$\xi = a\psi \tag{6.2.3}$$

Since σ is available as a function of ψ and not as a function of ξ, we need to apply the *chain rule* in eq. (6.2.2) in order to find a readily derivable expression for $\tan\alpha$, namely,

$$\tan\alpha = \frac{d\sigma}{d\psi}\frac{d\psi}{d\xi}$$

which, by virtue of eq. (6.2.3) leads to

$$\tan\alpha = \frac{1}{a}\sigma'(\psi) \tag{6.2.4a}$$

or, in terms of a normal motion $\tau(x)$, introduced in Section 2.2,

$$\tan\alpha = \frac{h}{a\,\Delta\psi}\tau'(x), \qquad 0 \le x \le 1 \tag{6.2.4b}$$

Next we impose the constraint that the absolute value of the pressure angle does not exceed a given bound α_M. This constraint leads to

$$\frac{h}{a\,\Delta\psi}|\,\tau'(x)\,| \le \tan\alpha_M$$

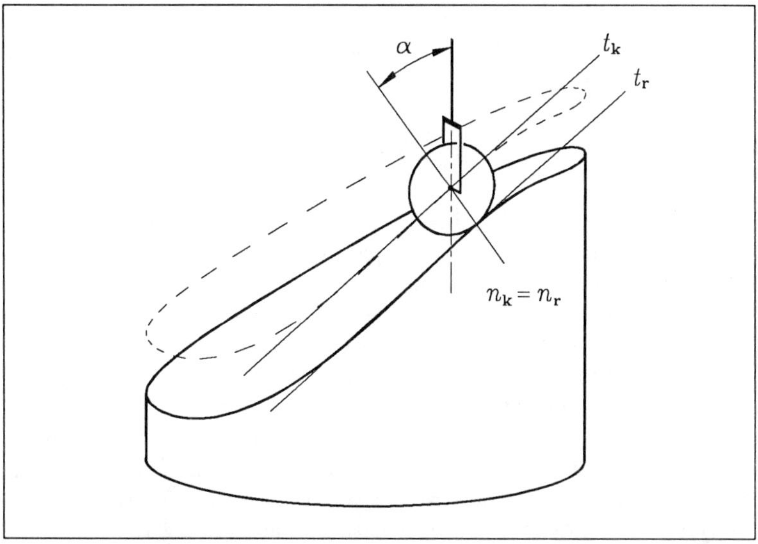

Fig. 6.2.6. Pressure angle for the pitch curve.

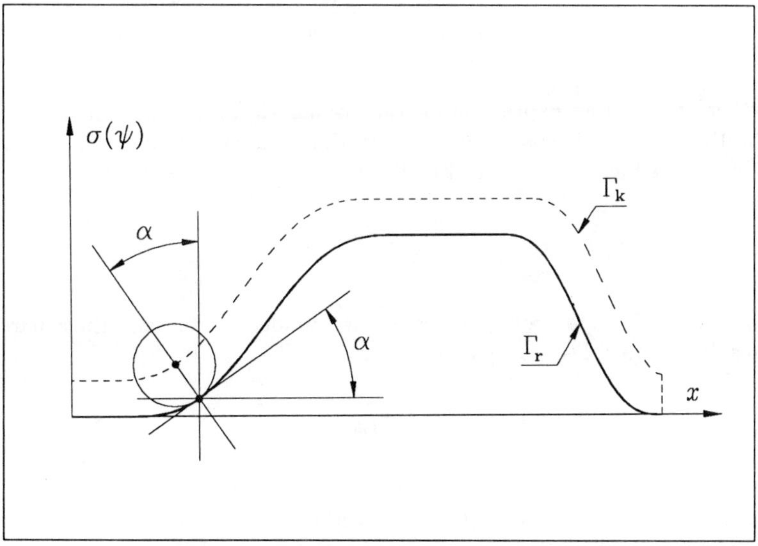

Fig. 6.2.7. Pressure angle of a cylindrical cam.

If we now let τ'_M denote the maximum value of $|\tau'(x)|$, the foregoing bound provides the minimum value of a/h that will keep the pressure angle within the prescribed bounds, namely,

$$\frac{a}{h} = \frac{\tau'_M}{\Delta\psi \tan\alpha_M} \tag{6.2.5}$$

which shows that the ratio a/h is directly proportional to the maximum value of the slope of the normal motion and inversely proportional to both the phase length $\Delta\psi$ and the tangent of the maximum value of the pressure angle. This result provides very useful information, namely, given two normal motions, the one with the smaller maximum slope produces smaller cams. For example, if we want to choose between the 3-4-5 polynomial motion and the cycloidal motion, the former will yield smaller cams for the same rise and the same bound on the pressure angle.

Once the radius of the pitch cylinder has been determined, the radius b of the roller is found based on the radius of curvature of the *developed* pitch curve Γ_k, which coincides with $\sigma(\psi)$. Below we derive an expression for the curvature κ of $\sigma(\psi)$. To this end, we denote the arc length measured along this curve by λ and recall an alternate definition of curvature, namely, as the rate of change of the slope of the curve under study with respect to the arc length. Thus, we have

$$\kappa \equiv \frac{d\tan\alpha}{d\lambda} = \frac{1}{a}\frac{d\sigma'(\psi)}{d\lambda} = \frac{1}{a}\frac{d\sigma'(\psi)}{d\psi}\frac{d\psi}{d\lambda} = \frac{\sigma''\psi)}{a\,\lambda'(\psi)}$$

Further, we need an expression for the arc length $\lambda(\psi)$. This is readily derived from the Pithagorean Theorem applied to a right triangle of sides $d\xi$ and $d\sigma$, with hypotenuse $d\lambda$, which yields, from Fig. 6.2.8,

$$d\lambda^2 = d\xi^2 + d\sigma^2$$

or, if both sides of the foregoing expression are divided by $d\psi^2$ and the square roots of the resulting two sides are equated,

$$\lambda'(\psi) = \sqrt{\xi'^2(\psi) + \sigma'^2(\psi)}$$

which is analogous to expression (3.2.2). Furthermore, from eq. (6.2.3), $\xi'(\psi) = a$, which thus leads to the desired expression for $\lambda'(\psi)$, namely,

$$\lambda'(\psi) = \sqrt{a^2 + \sigma'^2(\psi)}$$

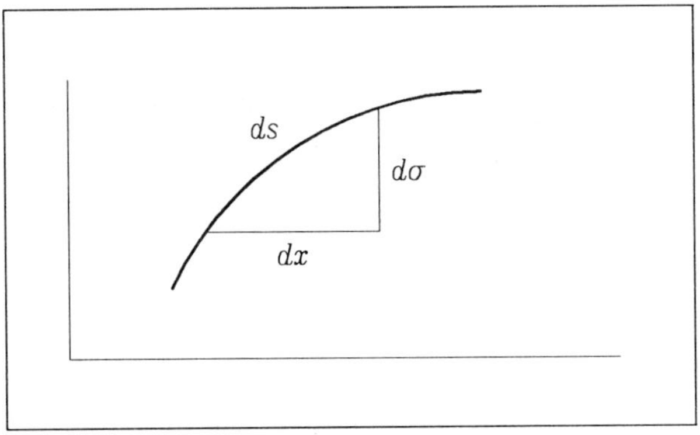

Fig. 6.2.8. Arc length.

Now the expression for $\lambda'(\psi)$ is substituted into the above expression for κ, thereby deriving the desired relation, namely,

$$\kappa = \frac{\sigma''(\psi)}{a\sqrt{a^2 + \sigma'^2(\psi)}} \tag{6.2.6a}$$

In order to avoid *undercutting*, we will dimension the radius of the roller not smaller than the minimum radius of curvature of σ. Now, since the radius of curvature is simply the reciprocal of the curvature, both items can attain positive and negative values, depending on the sign of $\sigma''(\psi)$, which is apparent from eq. (6.2.6a). Hence, the roller radius b will be chosen so as not to be smaller than the reciprocal of the maximum absolute value of the curvature, i.e.,

$$b \leq \frac{1}{\kappa_M} \tag{6.2.6b}$$

where

$$\kappa_M \equiv \max_{\psi}\{|\,\kappa(\psi)\,|\} \tag{6.2.6c}$$

Finding the maximum of $|\,\kappa(\psi)\,|$ is difficult because the absolute-value function is not differentiable at the origin. Thus, we will pursue instead the maximization of $\kappa^2(\psi)$, which is smooth enough. To this end, we evaluate $\kappa^2(\psi)$ from eq. (6.2.6a) as

$$\kappa^2(\psi) = \frac{\sigma''^2(\psi)}{a^2[\,a^2 + \sigma'^2(\psi)\,]} \tag{6.2.7a}$$

or, in terms of the normal motion $\tau(x)$,

$$\kappa^2(x) = \frac{\tau''^2(x)}{a^2(\Delta\psi)^2\left[\frac{a^2}{h^2}(\Delta\psi)^2 + \tau'^2(x)\right]} \equiv \frac{1}{(a\,\Delta\psi)^2}\frac{N(x)}{D(x)} \tag{6.2.7b}$$

Next, the derivative of κ^2 with respect to x is equated to zero, which leads to the *normality condition*

$$N'(x) - (a\,\Delta\psi)^2\,\kappa^2(x)\,D'(x) = 0 \tag{6.2.7c}$$

where

$$N'(x) = 2\,\tau''(x)\,\tau'''(x) \tag{6.2.7d}$$

$$D'(x) = 2\,\tau'(x)\,\tau''(x) \tag{6.2.7e}$$

Hence, the normality condition reduces to

$$\tau'''(x) - \frac{\tau''^2(x)}{\frac{a^2}{h^2}(\Delta\psi)^2 + \tau'^2(x)}\,\tau'(x) = 0$$

Now, the ratio a/h, as given by eq. (6.2.5), is substituted into the foregoing equation, which thus leads to

$$g(x) \equiv \left[\frac{\tau'^2_M}{\tan^2\alpha_M} + \tau'^2(x)\right]\tau'''(x) - \tau'(x)\,\tau''^2(x) = 0, \quad 0 \le x \le 1 \tag{6.2.7f}$$

thereby deriving the normality condition sought. Note that the roots of eq. (6.2.7f) determine the local extrema of $\kappa^2(x)$, and hence, those of $|\,\kappa(x)\,|$.

For arbitrary $\sigma(\psi)$, the normality condition can admit multiple roots, all of which represent local extrema or saddle points of r^2. From these, the smallest local minimum provides the *global minimum* r_0. Thus, the radius of the roller, b is constrained to be

$$b \le r_0 \tag{6.2.8a}$$

Note that, in the absence of other constraints, b can be made equal to r_0. Once the value of b is available, the cone angle β of the roller is determined as the acute angle obtained from

$$\beta = \tan^{-1}\left(\frac{b}{a}\right) \tag{6.2.8b}$$

thereby completing the optimum design. The sole geometric parameter that cannot be determined from purely kinematic constraints is the length of the contact line RS. This length is dictated by considerations of stress analysis and others like wear.

6.2.3 Example: Minimum-Size Cylindrical Cam with Translating Roller-Follower

Determine the cylindrical cam that will produce a rise h of its roller-follower of 100 mm in a cam rotation of 60°, for a maximum pressure angle of 30°. Design for *i)* cycloidal motion; and *ii)* 3-4-5 polynomial motion

Solution:

i) Cycloidal motion

In this case, from the result of Section 2.4, $\tau'_M = 2$. With this value and $\Delta\psi = \pi/3$, we obtain

$$\left(\frac{\tau'_M}{\tan\alpha_M}\right)^2 = (2\sqrt{3})^2 = 12$$

Moreover, from eq. (6.2.5),

$$a = \frac{100 \times 2 \times 3}{\pi \times 0.5773} = 330.83 \text{ mm}$$

Now, the curvature normality condition, eq. (6.2.7*f*), leads to

$$[\,12 + \tau'^2(x)\,]\,\tau'''(x) - \tau'(x)\,\tau''^2(x) = 0$$

Now, if $\tau(x)$ and its derivatives are substituted for the cycloidal motion, we obtain after simplifications

$$\cos^2 2\pi x - 14\cos 2\pi x + 1 = 0$$

which admits the roots

$$\cos 2\pi x_1 = 0.07180, \qquad \cos 2\pi x_2 = 13.93$$

and hence, only the first root leads to a real value of x. This yields

$$x_1 = 0.2386$$

Next, $N(x)$ and $D(x)$ are evaluated at x_1, and the computed numerical values are substituted into eq. (6.2.7*b*), which leads to

$$\kappa_{\max} = 0.05282\text{mm}^{-1}$$

Thus,

$$r_{\min} = \frac{1}{\kappa} = 189.3 \text{ mm}$$

and hence, the radius of the roller, b, can be bounded as

$$b \leq 189.3 \text{ mm}$$

which is a rather loose bound, for it allows a roller radius approximately 90% greater than the rise. If contact stresses are not a major concern, we could design a roller with $b = h = 100$ mm, which would yield a cone angle β as shown below:

$$\beta = \tan^{-1}\left(\frac{100}{330.83}\right) = 16.82°$$

ii) 3-4-5 Polynomial motion

For this case, from the result of Section 2.4, $\tau'_M = 1.875$. Substitution of this value into eq. (6.2.5) yields

$$a = \frac{100 \times 1.875 \times 3}{\pi \times 0.5773} = 310.15 \text{ mm}$$

and

$$\left(\frac{\tau'_M}{\tan\alpha_M}\right)^2 = 10.55$$

which leads to a curvature normality condition (6.2.7f) as shown below:

$$[\,10.55 + \tau'^2(x)\,]\,\tau'''(x) - \tau'(x)\,\tau''^2(x) = 0$$

Now, if $\tau(x)$ and its derivatives are substituted for the 3-4-5 polynomial motion, we obtain, after simplifications

$$P(x) \equiv 900\,x^4(x-1)^4(2x^2 - 2x - 1) + 10.55\,(6x^2 - 6x + 1) = 0$$

which admits two real roots in the interval $[\,0,\ 1\,]$ as shown in Fig. 6.2.9, namely,

$$x_1 = 0.1912, \qquad x_2 = 0.7670$$

Next, $N(x)$ and $D(x)$ are evaluated at x_1 and x_2, and the numerical values that are thus obtained are substituted into eq. (6.2.7b), which leads to two minima of the radius of curvature. These are

$$r_1 = 183.19 \text{ mm}, \qquad r_2 = 174.58 \text{ mm}$$

and hence, when we choose the lowest value, i.e., r_2, the radius of the roller is bounded as

$$b \leq 175.6 \text{ mm}$$

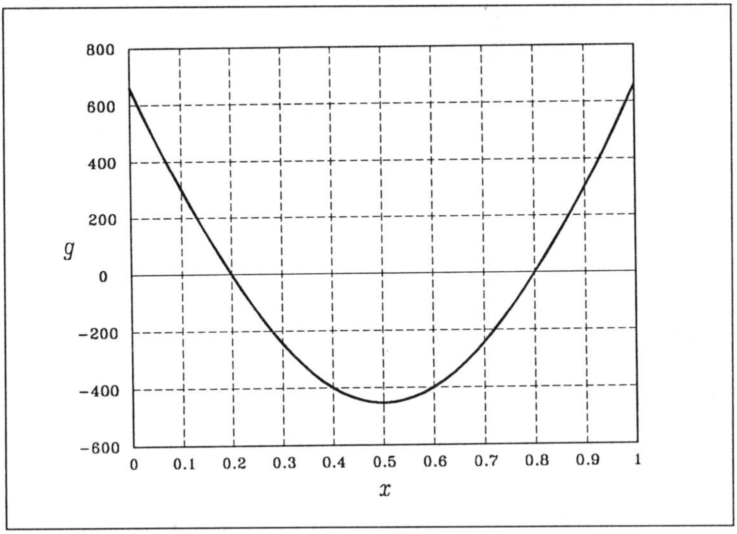

Fig. 6.2.9. Plot of $g(x)$, as given in eq. (6.2.7f), for 3-4-5 polynomial motion.

From the foregoing results, it is apparent that the polynomial motion produces lower bounds than the cycloidal motion, and hence, it is recommended when compared with the cycloidal motion.

6.2.4 Positive-Motion Cylindrical Cams with Translating Roller-Followers

As in other cam mechanisms, during the rise phase the load acts against the follower motion and hence produces a contact force that tends to keep the roller in contact with the cam cylinder. However, during the return phase, the load acts in a direction favoring the follower motion. Thus, it is possible that separation between the roller and the cam cylinder occurs. To prevent this, springs can be used, but these introduce lower resonant frequencies that can be detrimental to the dynamic operation of the mechanism. As an alternative, positive-motion cams can be designed as indicated below.

Positive contact is obtained between the two contacting surfaces by cutting a groove of suitable width on the cam cylinder, as indicated in Fig. 6.2.7, so as to allow a "perfect" fit of the roller in the groove. The outcome of this is that, whereas the roller will move under pure rolling with respect to one of the two tracks of the groove, sliding will unavoidably occur between the roller and the other track. This sliding will, in turn, introduce friction forces that will impinge upon the efficiency and the life span of the mechanism. To avoid friction forces, a clearance can be allowed between the groove and the roller, but this is not the best solution, because clearances provoke backlash along with all its negative effects.

As an alternative to the grooved track on the cam cylinder, the track can be *protruded* on the cylinder, as shown in Fig. 6.2.10. In this way, two rollers are mounted on the follower, thus producing a positive motion while avoiding sliding and backlash.

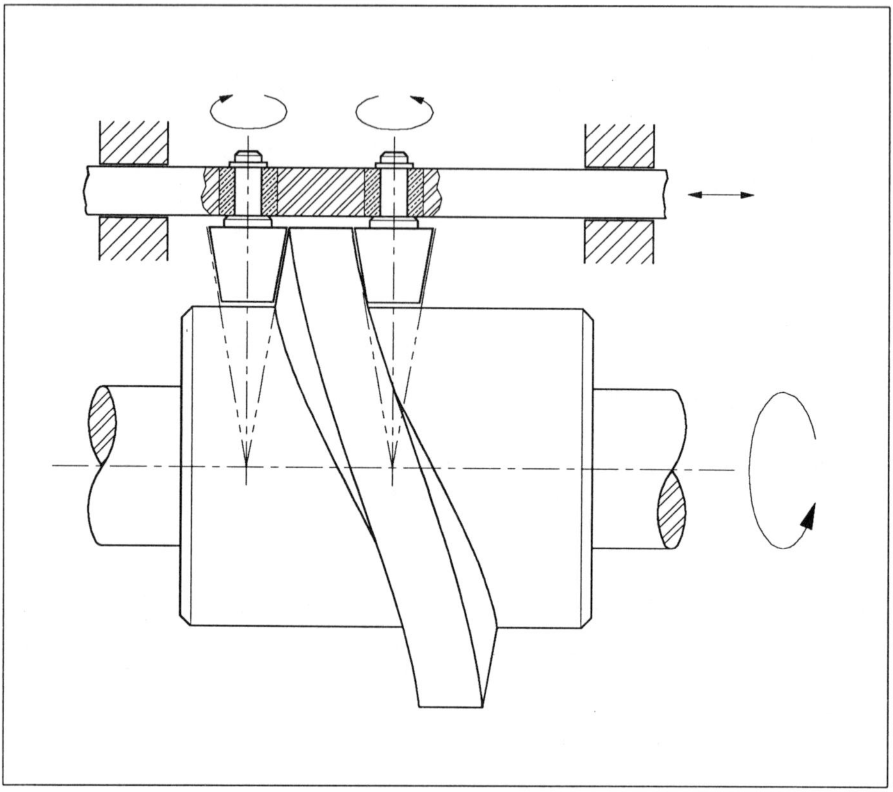

Fig. 6.2.10. Protruded track on a cylindrical cam.

Chapter 7

KINEMATIC ANALYSIS OF CAM MECHANISMS

7.1 Introduction

In this chapter, we discuss the kinematic analysis of cam mechanisms. That is, we determine the follower motion, given a cam profile described by the Cartesian coordinates of a discrete set of its points, its follower type, and all the parameters involved. We assume throughout that the cam profile is available as a discrete set of coordinate points $\{x(\psi_k),\ y(\psi_k)\}_1^m$. Moreover, we assume that we know the values of the first two derivatives of these Cartesian coordinates with respect to ψ, at the same ψ_k values. This is, indeed, the case if the cam profile was synthesized using the methods described in Chapters 4 and 5. If, on the other hand, the cam profile is available as an engineering blueprint, as a photograph or, even as a plate, then a database of point Cartesian coordinates can be constructed by suitably digitizing the given profile. However, the digitization of profile points yields only Cartesian coordinates but not their derivatives, which we can calculate by using a spline approximation of the cam profile, as outlined in Chapter 8. For translating followers, the follower output is given by the displacement $s(t)$, the velocity $\dot{s}(t)$, and the acceleration $\ddot{s}(t)$. For oscillating followers, the follower output is given by the angular displacement $\phi(t)$, the angular velocity $\dot{\phi}(t)$, and the angular acceleration $\ddot{\phi}(t)$. Once more, the basic approach for analysis will be computer-oriented rather than graphical. In Sections 7.2 and 7.3 we derive the analysis equations for translating and oscillating flat-face followers. In Section 7.4 we present the corresponding analysis for translating knife-edge followers, while Sections 7.5 and 7.6 deal with translating and oscillating roller-followers. Finally, in Section 7.7 we derive the analysis equations of cylindrical cams with translating roller-followers.

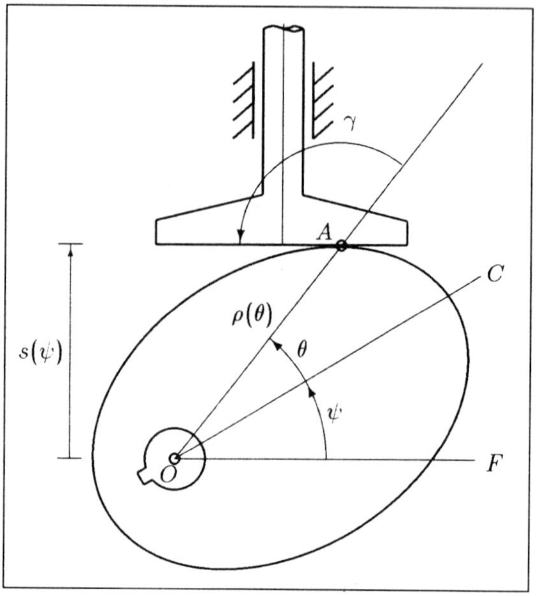

Fig. 7.2.1. Layout of a translating flat-face follower cam mechanism.

7.2 Translating Flat-Face Followers

The layout of this type of cam mechanism is shown in Fig. 7.2.1. In that figure, lines OF and OC are attached to the fixed frame and to the cam, respectively. The cam angle of rotation is denoted by ψ, while the follower output is indicated as $s(\psi)$. Moreover, ρ—segment $\overline{OA}$—and θ are the polar coordinates describing the cam profile.

7.2.1 Displacement Analysis

From the geometry of Fig. 7.2.1, we have

$$s(\psi) = \rho(\theta) \sin(\psi + \theta) \tag{7.2.1}$$

Upon expansion of the right-hand side of eq. (7.2.1), $s(\psi)$ becomes

$$s(\psi) = \rho(\theta) \cos\theta \sin\psi + \rho(\theta) \sin\theta \cos\psi \tag{7.2.2a}$$

or, in terms of the Cartesian coordinates of the cam profile,

$$s(\psi) = x(\psi) \sin\psi + y(\psi) \cos\psi \tag{7.2.2b}$$

which is the relation sought.

7.2.2 Velocity Analysis

Differentiation of the follower displacement function $s(\psi)$ with respect to time yields the follower velocity, i.e.,

$$\dot{s}(t) = s'(\psi)\dot{\psi} \tag{7.2.3a}$$

where the *chain rule* has been applied. Furthermore, $s'(\psi)$ is computed by differentiation of both sides of eq. (7.2.2*b*) with respect to ψ, which yields, as derived in eq. (4.2.8*b*),

$$s'(\psi) = x(\psi)\cos\psi - y(\psi)\sin\psi \tag{7.2.3b}$$

and hence,

$$\dot{s}(\psi) = [\,x(\psi)\cos\psi - y(\psi)\sin\psi\,]\,\dot{\psi} \tag{7.2.3c}$$

which is the relation sought.

7.2.3 Acceleration Analysis

Upon differentiation of both sides of eq. (7.2.3*a*) with respect to time, and applying the *chain rule*, the acceleration of the follower is determined as:

$$\ddot{s}(t) = s'(\psi)\,\ddot{\psi} + s''(\psi)\,\dot{\psi}^2(t) \tag{7.2.4}$$

where $s''(\psi)$ is obtained from eq. (4.2.28*d*), which is reproduced below for quick reference:

$$\mathbf{r}'(\psi) = \mathbf{E}\,\mathbf{r}(\psi) + \mathbf{Q}(\psi)\,\mathbf{p}(\psi) \tag{7.2.5}$$

The first component of the foregoing equation yields

$$s''(\psi) = -s(\psi) + x'(\psi)\cos\psi - y'(\psi)\sin\psi \tag{7.2.6}$$

and hence, the follower acceleration takes on the form

$$\begin{aligned}\ddot{s}(\psi) = {} & [x(\psi)\cos\psi - y(\psi)\sin\psi]\ddot{\psi} + \\ & \{[-x(\psi) - y'(\psi)]\sin\psi + [-y(\psi) + x'(\psi)]\cos\psi\}\dot{\psi}^2\end{aligned} \tag{7.2.7}$$

which is the desired expression.

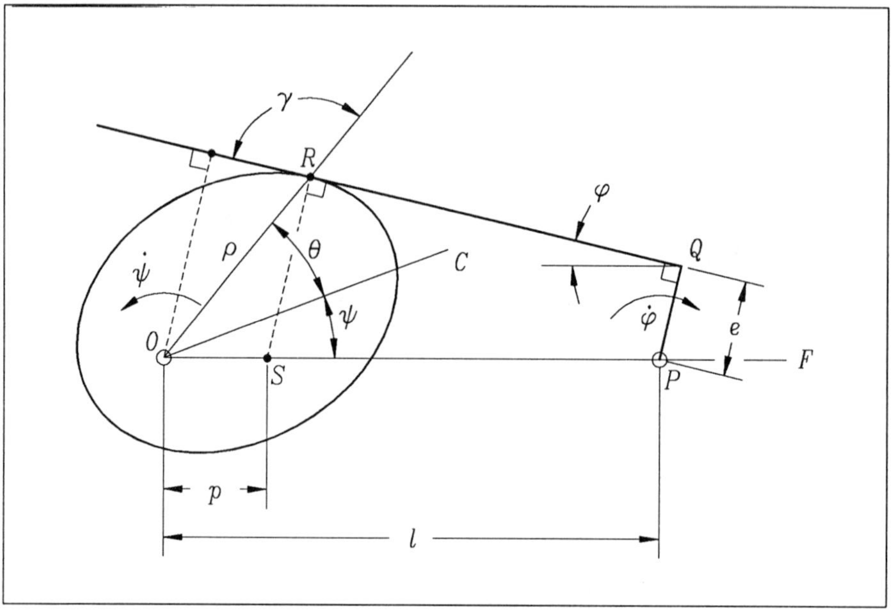

Fig. 7.3.1. Layout of a disk cam with oscillating flat-face follower.

7.3 Oscillating Flat-Face Followers

Let us consider the cam-mechanism layout shown in Fig. 7.3.1. The cam profile is assumed to be given byits Cartesian coordinates $x(\psi)$, $y(\psi)$.

7.3.1 Displacement Analysis

We can rewrite eqs. (4.3.11) and (4.3.12) in the form

$$\ell\,\frac{\phi'\cos\phi}{1+\phi'} = x\cos(\psi+\phi) - y\sin(\psi+\phi) \tag{7.3.1a}$$

$$\ell\sin\phi + e = x\sin(\psi+\phi) + y\cos(\psi+\phi) \tag{7.3.1b}$$

Upon expansion of terms on the right-hand side, in vector form, we obtain

$$\begin{bmatrix} \ell\,\frac{\phi'\cos\phi}{1+\phi'} \\ \ell\sin\phi + e \end{bmatrix} = \begin{bmatrix} \cos\phi & -\sin\phi \\ \sin\phi & \cos\phi \end{bmatrix} \begin{bmatrix} x\cos\psi - y\sin\psi \\ x\sin\psi + y\cos\psi \end{bmatrix}$$

from which we can readily transfer the terms in ϕ appearing on the right-hand side to the left-hand side. To this end, we exploit the orthogonality of the matrix coefficient on the right-hand side, thereby deriving

$$\begin{bmatrix} \cos\phi & \sin\phi \\ -\sin\phi & \cos\phi \end{bmatrix} \begin{bmatrix} \ell \frac{\phi' \cos\phi}{1+\phi'} \\ \ell \sin\phi + e \end{bmatrix} = \begin{bmatrix} x\cos\psi - y\sin\psi \\ x\sin\psi + y\cos\psi \end{bmatrix}$$

Upon performing the product on the left-hand side, two equations in ϕ and ϕ' are derived, namely,

$$\ell \frac{\phi' + \sin^2\phi}{1+\phi'} + e\sin\phi = a \tag{7.3.2a}$$

$$\ell \frac{\cos\phi \sin\phi}{1+\phi'} + e\cos\phi = b \tag{7.3.2b}$$

with a and b defined as

$$a \equiv x\cos\psi - y\sin\psi \tag{7.3.2c}$$

$$b \equiv x\sin\psi + y\cos\psi \tag{7.3.2d}$$

Now, we solve for ϕ' from each of the two equations (7.3.2*a* & *b*), thereby deriving two independent expressions for this variable, i.e.,

$$\phi' = \frac{a - \ell\sin^2\phi - e\sin\phi}{\ell + e\sin\phi - a} \tag{7.3.3a}$$

$$\phi' = \frac{b - \ell\cos\phi\sin\phi - e\cos\phi}{e\cos\phi - b} \tag{7.3.3b}$$

from which we readily obtain

$$\frac{a - \ell\sin^2\phi - e\sin\phi}{\ell + e\sin\phi - a} = \frac{b - \ell\cos\phi\sin\phi - e\cos\phi}{e\cos\phi - b}$$

and hence,

$$(a - \ell\sin^2\phi - e\sin\phi)(e\cos\phi - b) = (\ell + e\sin\phi - a)(b - \ell\cos\phi\sin\phi - e\cos\phi)$$

After simplifications, the above equation reduces to

$$\cos\phi\,[\,b\cos\phi - (\ell - a)\sin\phi - e\,] = 0$$

Thus, for $\cos\phi \neq 0$, the last equation yields

$$b\cos\phi - (\ell - a)\sin\phi - e = 0 \tag{7.3.4}$$

which is a trigonometric equation allowing us to calculate ϕ for a given value of ψ. In order to solve for ϕ from the above equation, the usual trigonometric identities are introduced, namely,

$$\cos\phi \equiv \frac{1-T^2}{1+T^2}, \quad \sin\phi \equiv \frac{2T}{1+T^2}, \quad T \equiv \tan\left(\frac{\phi}{2}\right) \tag{7.3.5}$$

thereby deriving a quadratic equation in T, as shown below

$$(b+e)T^2 + 2(\ell - a)T - (b-e) = 0 \tag{7.3.6a}$$

The two roots of eq. (7.3.6a) are

$$T_{1,2} = \frac{a - \ell \pm \sqrt{R}}{b+e} \tag{7.3.6b}$$

the radical R being defined as

$$R \equiv (a-\ell)^2 + (b+e)(b-e)$$

Upon expansion, R reduces to

$$R = x^2(\psi) + y^2(\psi) + \ell^2 - e^2 - 2\ell[x(\psi)\cos\psi - y(\psi)\sin\psi] \tag{7.3.6c}$$

and hence, $T_{1,2}$ can be given explicitly as functions of ψ, namely,

$$T_{1,2} = \frac{x(\psi)\cos\psi - y(\psi)\sin\psi - \ell \pm \sqrt{R(\psi)}}{x(\psi)\sin\psi + y(\psi)\cos\psi + e} \tag{7.3.6d}$$

and the two arising values of ϕ are thus calculated as

$$\phi_i(\psi) = 2\tan^{-1}[T_i(\psi)], \quad i = 1,2 \tag{7.3.6e}$$

In summary, then, for every position of the cam plate, defined by ψ, two positions of the follower, given by ϕ_1 and ϕ_2, are possible. This result is in agreement with the geometry of the mechanism.

7.3.2 Velocity Analysis

Upon differentiation of the angular displacement, ϕ, with respect to time, and applying the *chain rule*, we can compute the angular velocity of the follower as

$$\dot{\phi}(t) = \phi'(\psi) \tag{7.3.7}$$

where $\phi'(\psi)$ can be calculated from either of eqs. (7.3.3a & b). We will obviously choose the more *robust* of the two formulas. Here, robustness is understood in the numerical sense. Thus, of the two above-mentioned formulas, the more robust is that producing the smaller round-off error. This is the formula having the denominator with the largest absolute value.

7.3.3 Acceleration Analysis

Differentiation of both sides of eq. (7.3.7) with respect to time gives the angular acceleration of the follower, namely,

$$\ddot{\phi}(t) = \phi'(\psi)\,\ddot{\psi} + \phi''(\psi)\,\dot{\psi}^2 \tag{7.3.8}$$

where the *chain rule* has been applied. Now, we need to compute $\phi''(\psi)$, which is done by straightforward differentiation of either of eqs. (7.3.2a) and (7.3.2b), rewritten in the form

$$\ell(\phi' + \sin^2\phi) + (1+\phi')(e\sin\phi - a) = 0 \tag{7.3.9a}$$

$$\ell\cos\phi\sin\phi + (1+\phi')(e\cos\phi - b) = 0 \tag{7.3.9b}$$

Upon differentiation of both sides of each of the two foregoing equations with respect to ψ, and solving the arising equations for ϕ'', we obtain, correspondingly,

$$\phi''(\psi) = -\frac{2\ell\cos\phi\sin\phi + (1+\phi')(e\phi'\cos\phi - a')}{\ell + e\sin\phi - a} \tag{7.3.10a}$$

$$\phi''(\psi) = \frac{\ell\phi'(\sin^2\phi - \cos^2\phi) + (1 + \phi'\sin\phi + b')}{e\cos\phi - b} \tag{7.3.10b}$$

with $a'(\psi)$ and $b'(\psi)$ calculated from eqs. (7.3.2c & d), namely,

$$a'(\psi) = [x'(\psi) - y]\cos\psi - [y'(\psi) + x]\sin\psi \tag{7.3.11a}$$

$$b'(\psi) = [y'(\psi) + x]\cos\psi + [x'(\psi) - y]\sin\psi \tag{7.3.11b}$$

From the foregoing expressions, then, it is apparent that both $\phi'(\psi)$ and $\phi''(\psi)$ have the same denominators. As a consequence, the more robust formula for $\phi'(\psi)$ will be likewise the more robust formula for $\phi''(\psi)$.

7.4 Translating Knife-Edge Followers

Let us consider the cam-follower mechanism shown in Fig. 7.4.1. Line OF is attached to the fixed frame, whereas line OC is attached to the cam plate. Moreover, ψ is the cam angle of rotation, whereas the follower offset is denoted by e. We will assume again that the cam profile is available from its Cartesian coordinates $x(\psi)$ and $y(\psi)$.

7.4.1 Displacement Analysis

From the geometry of Fig. 7.4.1, we can readily derive the relation

$$s(\psi) = \rho(\theta) \sin(\psi + \theta) \tag{7.4.1}$$

Now, we recall that

$$x(\psi) = \rho(\theta) \cos\theta, \quad y(\psi) = \rho(\theta) \sin\theta \tag{7.4.2}$$

where $\theta = \theta(\psi)$,

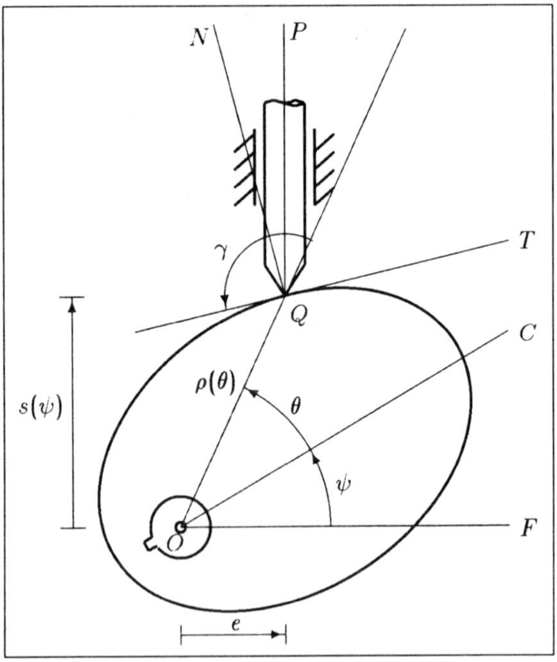

Fig. 7.4.1. Geometry of a cam mechanism with a translating knife-edge follower

Upon the expansion of the right-hand side of eq. (7.4.1) and the introduction of eq.(7.4.2), it is apparent that

$$s(\psi) = x(\psi)\sin\psi + y(\psi)\cos\psi \tag{7.4.3}$$

which is the desired relation.

7.4.2 Velocity Analysis

In order to determine the velocity of the follower, we recall eq. (5.2.40b), namely,

$$\mathbf{r}'(\psi) = \mathbf{Q}(\psi)\,\mathbf{p}'(\psi) + \mathbf{E}\,\mathbf{r}$$

i.e.,

$$\begin{bmatrix} 0 \\ s'(\psi) \end{bmatrix} = \begin{bmatrix} \cos\psi & -\sin\psi \\ \sin\psi & \cos\psi \end{bmatrix} \begin{bmatrix} x'(\psi) \\ y'(\psi) \end{bmatrix} + \begin{bmatrix} -s(\psi) \\ e \end{bmatrix}$$

from which we readily obtain

$$s'(\psi) = x'(\psi)\sin\psi + y'(\psi)\cos\psi + e \tag{7.4.4}$$

thereby deriving the follower velocity in the form

$$\dot{s}(t) = s'(\psi)\,\dot{\psi}(t) \tag{7.4.5}$$

The velocity analysis of radial followers is just a particular case of the more general offset followers. Hence, $s'(\psi)$ for radial followers is calculated from eq. (7.4.4) by setting $e = 0$.

7.4.3 Acceleration Analysis

The acceleration $\ddot{s}(t)$ is calculated as in eq. (7.2.4), which requires the calculation of $s''(\psi)$. This is readily calculated by recalling eq. (5.2.40c), namely,

$$\mathbf{r}''(\psi) = \mathbf{Q}(\psi)\,\mathbf{p}''(\psi) + 2\,\mathbf{E}\,\mathbf{r}'(\psi) + \mathbf{r}$$

that is,

$$\begin{bmatrix} 0 \\ s''(\psi) \end{bmatrix} = \begin{bmatrix} \cos\psi & -\sin\psi \\ \sin\psi & \cos\psi \end{bmatrix} \begin{bmatrix} x''(\psi) \\ y''(\psi) \end{bmatrix} + 2\begin{bmatrix} -s'(\psi) \\ 0 \end{bmatrix} + \begin{bmatrix} 0 \\ s'(\psi) \end{bmatrix}$$

The second component of $\mathbf{r}''(\psi)$ of the above expression yields

$$s''(t) = x''(\psi)\sin\psi + y''(\psi)\cos\psi + s'(\psi) \tag{7.4.6}$$

For the case of radial followers, the acceleration of the follower is determined from eq. (7.4.6) as well, except that now e is set equal to 0 when calculating $s'(\psi)$.

7.5 Translating Roller-Followers

Fig. 7.5.1 shows a layout of a cam mechanism with a translating roller-follower. The radius of the follower is denoted by a, and all the parameters and variables shown in that figure have the same meaning as in the previous section.

The displacement, velocity, and acceleration analyses of this type of mechanism are reduced to those of knife-edge followers. Indeed, the follower displacement of the mechanisms under study is identical to that of the knife-edge follower driven by a cam whose profile is given by the *pitch curve* of the given mechanism. It is recalled here that the pitch curve is the trajectory of the center of the roller on the cam plate.

If $\mathbf{r}_k$ and $\mathbf{r}_r$ denote the position vectors of the *corresponding points* on the pitch curve and the cam profile of the mechanism at hand, as indicated in Fig. 5.2.9, then, the relation between those position vectors, as given by eq. (5.2.37), leads to

$$\mathbf{r}_k = \mathbf{r}_r - a\mathbf{e}_n \tag{7.5.1a}$$

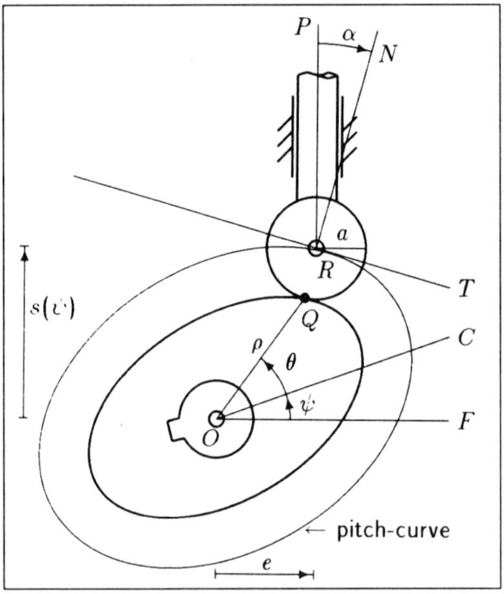

Fig. 7.5.1. Geometry of a cam mechanism with a translating roller-follower.

where

$$\mathbf{r}_k = \begin{bmatrix} x_k(\psi) \\ y_k(\psi) \end{bmatrix}, \quad \mathbf{r}_r = \begin{bmatrix} x(\psi) \\ y(\psi) \end{bmatrix} \tag{7.5.1b}$$

Moreover, if κ denotes the curvature of the given profile, then

$$\mathbf{e}_n = \operatorname{sgn}(\kappa)\,\mathbf{E}\,\mathbf{e}_t \tag{7.5.1c}$$

with $\mathbf{E}$ and $\mathbf{e}_t$ defined as in Section 3.2, namely,

$$\mathbf{E} = \begin{bmatrix} 0 & -1 \\ 1 & 0 \end{bmatrix}, \quad \mathbf{e}_t \equiv \frac{1}{\sqrt{x'^2(\psi) + y'^2(\psi)}} \begin{bmatrix} x'(\psi) \\ y'(\psi) \end{bmatrix} \tag{7.5.1d}$$

Therefore,

$$\mathbf{r}_k = \frac{1}{\sqrt{x'^2(\psi) + y'^2(\psi)}} \begin{bmatrix} x(\psi)\sqrt{x'^2(\psi) + y'^2(\psi)} + a\operatorname{sgn}(\kappa) y'(\psi) \\ y(\psi)\sqrt{x'^2(\psi) + y'^2(\psi)} - a\operatorname{sgn}(\kappa) x'(\psi) \end{bmatrix} \tag{7.5.2}$$

Once the pitch curve is produced using eq. (7.5.2), the derivatives of its coordinates, $x_k'(\psi)$, $y_k'(\psi)$, $x_k''(\psi)$, and $y_k''(\psi)$, are calculated using a spline approximation, as discussed in Chapter 8. With this information, the kinematic analyses of Section 7.4 are straightforward.

7.6 Oscillating Roller-Followers

The geometry of this type of cam mechanism is shown in Fig. 7.6.1. In that figure, lines T and N are the tangent and the normal of the pitch curve, respectively. Moreover, the length of the follower arm is e, and the radius of the roller is a.

7.6.1 Displacement Analysis

Here we recall eqs. (5.3.5a & b), and rewrite them in the form

$$\ell - e\cos\phi = x(\psi)\cos\psi - y(\psi)\sin\psi \tag{7.6.1a}$$

$$e\sin\phi = x(\psi)\sin\psi + y(\psi)\cos\psi \tag{7.6.1b}$$

from which,

$$\cos\phi = \frac{\ell - x(\psi)\cos\psi + y(\psi)\sin\psi}{e} \tag{7.6.2a}$$

$$\sin\phi = \frac{x(\psi)\sin\psi + y(\psi)\cos\psi}{e} \tag{7.6.2b}$$

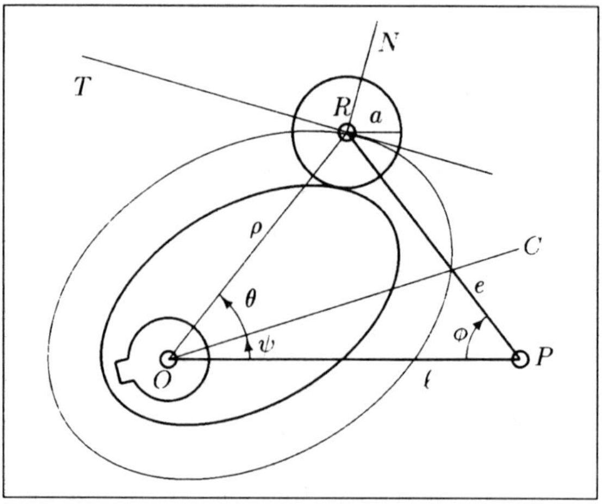

Fig. 7.6.1. Cam mechanism with oscillating roller-follower

From eqs. (7.6.2a & b), ϕ can be determined uniquely as

$$\phi = \tan^{-1}\left[\frac{x(\psi)\sin\psi + y(\psi)\cos\psi}{\ell - x(\psi)\cos\psi + y(\psi)\sin\psi}\right] \tag{7.6.3}$$

which thus provides ϕ as a function of ψ.

7.6.2 Velocity Analysis

Differentiation of each of eqs. (7.6.1a & b) with respect to ψ yields

$$e\phi'\sin\phi = [x'(\psi) - y(\psi)]\cos\psi - [y'(\psi) + x(\psi)]\sin\psi \tag{7.6.4a}$$

$$e\phi'\cos\phi = [y'(\psi) + x(\psi)]\cos\psi + [x'(\psi) - y(\psi)]\sin\psi \tag{7.6.4b}$$

from which two expressions for $\phi'(\psi)$ can be derived. Which equation to use depends on the value of ϕ but, as a rule, one would use the more robust of these two expressions. Here, the more robust expression will be that yielding the coefficient of ϕ' with the larger absolute value. Rather than merely choose the expression with this property, we can derive a single expression that is robust for any value of ϕ. This expression can be derived if both sides of eq. (7.6.4a) are multiplied by $\sin\phi$, and both sides of eq. (7.6.4b) are multiplied by $\cos\phi$, and the resulting equations are then added, which yields

$$\phi'(\psi) = \frac{a(\psi)\sin\phi + b(\psi)\cos\phi}{e} \tag{7.6.5a}$$

with $a(\psi)$ and $b(\psi)$ defined as

$$a(\psi) \equiv [x'(\psi) - y(\psi)]\cos\psi - [y'(\psi) + x(\psi)]\sin\psi \tag{7.6.5b}$$

$$b(\psi) \equiv [y'(\psi) + x(\psi)]\cos\psi + [x'(\psi) - y(\psi)]\sin\psi \tag{7.6.5c}$$

Now, it is a simpler matter to calculate the angular velocity $\dot{\phi}(t)$ as

$$\dot{\phi}(t) = \phi'(\psi)\dot{\psi}(t) \tag{7.6.6}$$

7.6.3 Acceleration Analysis

The acceleration analysis is then straightforward, for all we need is to differentiate both sides of eq. (7.6.5a) with respect to ψ, thereby obtaining

$$\phi''(\psi) = \frac{[a'(\psi) - b(\psi)\phi'(\psi)]\sin\phi + [a(\psi)\phi'(\psi) + b'(\psi)]\cos\phi}{e} \tag{7.6.7a}$$

where $a'(\psi)$ and $b'(\psi)$ are given as

$$a'(\psi) \equiv (x'' - 2y' - x)\cos\psi - (y'' + 2x' - y)\sin\psi \tag{7.6.7b}$$

$$b'(\psi) \equiv (y'' + 2x' - y)\cos\psi + (x'' - 2y' - x)\sin\psi \tag{7.6.7c}$$

where the argument ψ has been omitted in x', x'', y', and y'' for brevity.

Now, the angular acceleration $\ddot{\phi}(t)$ is calculated simply as

$$\ddot{\phi}(t) = \phi'(\psi)\ddot{\psi}(t) + \phi''(\psi)\dot{\psi}^2(t) \tag{7.6.8}$$

thereby completing the full kinematic analysis.

7.7 Cylindrical Cam with Translating Roller-Follower

The *developed* curves of the knife-edge and the roller-follower are shown in Fig. 7.7.1. In that figure, b is the roller radius, $\sigma_r(\xi)$ defines the profile of the roller track, with $\xi = a\psi$, and a is the radius of the pitch cylinder. Moreover, $\sigma_k(\xi)$ denotes the displacement of the roller center, which is to be determined from knowledge of $\sigma_r(\xi)$, $\sigma_r'(\xi)$, and $\sigma_r''(\xi)$, along with the cam parameters a and b. From Fig. 7.7.1,

$$\sigma_k(\psi) = \sigma_r(\xi) + a\,\cos\alpha \tag{7.7.1}$$

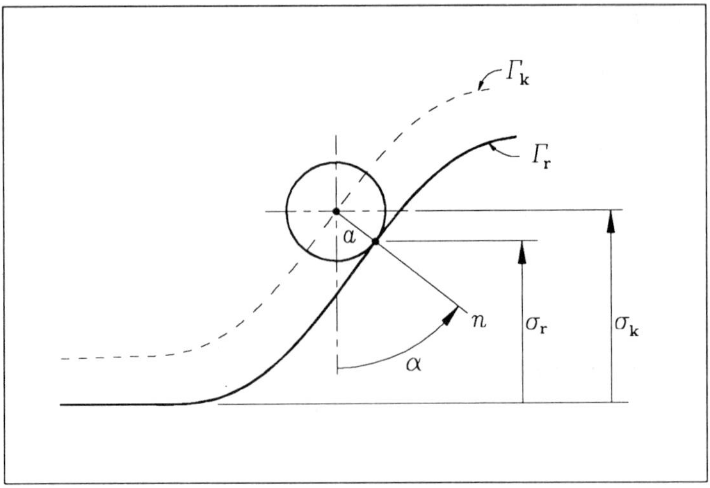

Fig. 7.7.1. *Developed* pitch-curve and cam profile of a cylindrical cam.

where the pressure angle α is given through its tangent, as derived in eq. (6.2.2), namely,

$$\tan\alpha = \sigma_r'(\xi) \tag{7.7.2}$$

Thus,

$$\cos\alpha = \frac{1}{\sqrt{1+\tan^2\alpha}} = \frac{1}{\sqrt{1+\sigma'^2_r(\xi)}} \tag{7.7.3}$$

Substitution of eq. (7.7.3) into eq. (7.7.1) yields

$$\sigma_k(\psi) = \sigma_r(\xi) + \frac{a}{\sqrt{1+\sigma'^2_r(\xi)}} \tag{7.7.4}$$

which is the desired expression for the follower displacement.

For the velocity analysis, $\sigma_k'(\xi)$ will be needed. This derivative is obtained by differentiation of both sides of eq. (7.7.4) with respect to ξ, which yields

$$\sigma_k'(\xi) = \sigma_r'(\xi) - \frac{a\sigma_r'(\xi)\,\sigma_r''(\xi)}{[1+\sigma'^2_r(\xi)]^{3/2}} \tag{7.7.5}$$

Likewise, the acceleration of the follower requires the calculation of $\sigma_k''(\xi)$, which is derived below, upon differentiation of both sides of eq. (7.7.5) with respect to ξ, namely,

$$\sigma_k''(\xi) = \sigma_r''(\xi) - a\,\frac{\sigma''^2_r(\xi) - \sigma_r'(\xi)\sigma_r'''(\xi) - 2\sigma'^2_r(\xi)\sigma''^2_r(\xi) - \sigma_r'(\xi)\sigma_r'''(\xi)}{[1+\sigma'^2_r(\xi)]^{5/2}} \tag{7.7.6}$$

Thus, the velocity of the follower, $\dot{\sigma}_k(t)$, is derived as

$$\dot{\sigma}(t) = \sigma_k'(\xi)\,\dot{\xi}(t)$$

where, from eq. (6.2.3),

$$\dot{\xi}(t) = a\,\dot{\psi}(t)$$

and hence,

$$\dot{\sigma}_k(t) = a\,\sigma_k'(\xi)\,\dot{\psi}(t) \tag{7.7.7}$$

Likewise,

$$\ddot{\sigma}_k(t) = \sigma_k'(\xi)\,\ddot{\xi}(t) + \sigma_k''(\xi)\,\dot{\xi}^2(t)$$

and hence,

$$\ddot{\sigma}_k(t) = a\sigma_k'(\xi)\ddot{\psi}(t) + a^2\sigma_k''(\xi)\dot{\psi}^2(t) \tag{7.7.8}$$

thereby completing the kinematic analysis.

7.8 Summary

From inspection of the acceleration expressions for flat-face follower cam mechanisms with both translating and oscillating followers, eqs. (7.2.7) and (7.3.10*a* & *b*), it is apparent that these mechanisms are *integrators* in the sense of systems dynamics (Cannon, 1967). This means that the follower acceleration depends only on the first derivatives of the Cartesian coordinates defining the cam profile. Thus, if the cam profile for this type of cam is approximated, for manufacture, as a succession of tangent circles of variable radius, the cam coordinates $x(\psi)$ and $y(\psi)$ are continuous up to first-order derivatives, but their second derivatives with respect to ψ are discontinuous at the points of tangency. These discontinuities will not affect the follower acceleration, because of the integrating effect of these mechanisms.

On the other hand, from expressions (7.4.6) and (7.6.6*a*–*c*), it is apparent that the follower acceleration of disk cam mechanims with roller-followers depends upon the second-order derivatives of the cam profile with respect to ψ. Hence, any discontinuities in these coordinates will invariably appear in the follower acceleration, and special care must be taken when designing and cutting the cam profile of these mechanims. Note that acceleration discontinuities in the follower acceleration will induce inertia force discontinuities, which will affect the overall dynamic response of the machine.

Finally, inspection of eq. (7.7.6) makes apparent that cylindrical cams with roller-followers are *differentiators*. That is, the acceleration of the follower is influenced by up to third-order derivatives of the roller track with respect to ξ—or ψ, for that matter. Hence, if cubic splines are used to design and manufacture the roller track, then a large number of supporting points should be used, in order to avoid large discontinuities of the third derivative of the spline. In fact, discontinuities in that derivative with respect to ξ will unavoidably produce discontinuities in the acceleration of the follower, and hence in the inertia forces acting on the overall machine. The numerical aspects of cam profile approximation and calculation are discussed in detail in Chapter 8.

Chapter 8 COMPUTATION OF THE GEOMETRIC PROPERTIES OF CAMS

8.1 Introduction

In this chapter we discuss the accurate and fast computation of the first three moments—volume, vector first moment, and inertia matrix—of regions bounded by 2-D contours and 3-D surfaces. The approaches that have been reported in this regard vary in nature, although most of them refer to a discrete integration, namely, a summation of integrals over a set of subregions (Lee and Requicha, 1982a and b; Woodwark and Quinlan, 1982). Some methods resorting to the Gauss Divergence Theorem (GDT) have also been reported (Messner and Taylor, 1980; Miles and Tough, 1983; Angeles, 1983; Angeles et al. 1990; Lien and Kajiya, 1984; Ota et al., 1985). The latter are especially attractive because they allow the reduction of area or volume integrals to either line or surface integrals. Whereas formulas for the transformation of integrals defining the volume and the first moment of solid regions bounded by closed surfaces into surface integrals have been available for some time (Brand, 1965; Hurley, 1981), computer-oriented algorithms making use of such formulas are somewhat recent. In fact, Messner and Taylor (1980) are recognized as the first to make full use of the GDT in this context. Lien and Kajiya (1984) proposed simple formulas based on the GDT that allow an exact evaluation of the moments of arbitrary polyhedra. Although the formulas they propose are quite general, for they allow the computation of moments of any order—see Section 8.3 for a definition of these—, they are limited to polyhedra; on the other hand, they do not include explicit formulas for the functions required to evaluate the surface integrals involved. Ota et al. (1985) proposed more explicit formulas making use of the GDT. These formulas, however, do not supply all components of the inertia matrix but only its projections onto one given axis, thereby excluding products of inertia. In this regard, Messner and Taylor (1980) included products of inertia, the computation of the surface integrals being done through a quadrature rule applied to triangular elements of the surface polygons. Timmer and Stern (1980) computed the moments of solids bounded by surfaces which are approximated by parameterized patches rather than polygons. Using the GDT and Green's Theorem, they reduced the problem to evaluating line integrals over the contours of the surface patches. The line integrals are computed by approximate integration formulas. These formulas provide the required moments, excluding off-diagonal entries of the inertia matrix—the products of inertia. Angeles et al. (1988) proposed a method based on the GDT, for the computation of moments of planar and axially symmetric regions, given their spline approximation of the boundary. The method is general enough to supply all the components of the inertia matrix.

In Section 8.2, periodic parametric cubic spline functions are recalled. The general formulas derived in Angeles et al. (1988), aimed at the computation of the kth moment of bounded ν-dimensional regions, are discussed in Section 8.3. In Section 8.4, the aforementioned formulas are applied to the computation of the area, centroid coordinates, and the matrix second moment of two-dimensional regions. The boundaries of these regions can be approximated either by piecewise linear functions or by spline functions. Section 8.5 introduces practical simple formulas for the computation of the volume, centroid coordinates, and inertia matrix of solid regions. Finally, Section 8.6 includes examples of the computation of the geometric properties of cams, based on their boundary approximations. The latter are assumed to be represented by periodic parametric cubic splines, which are discussed below.

8.2 Periodic Parametric Cubic Splines

In this section, we focus on the representation of the closed contours of the planar cam mechanisms which were studied in Chapters 4 and 5. We show that these contours can be represented conveniently with periodic cubic splines.

As discussed in Chapters 4 and 5, the cam profiles of interest are assumed to be closed smooth curves of the C^2 type. Moreover, functions describing the Cartesian coordinates of a cam profile, i.e., $x = x(\psi)$ and $y = y(\psi)$, are also periodic functions of ψ. In the discussion below we introduce an alternative parameter, p, yet to be defined, and regard x and y as functions $x(p)$ and $y(p)$.

Let $(x_i,\ y_i)$, where $x_i \equiv x(\psi_i)$ and $y_i \equiv y(\psi_i)$, for $i = 1,\ldots,n$, be the set of generated points resulting from the cam synthesis equations derived in Chapters 4 and 5. Since we need a closed smooth curve, the pertinent periodic boundary conditions must be satisfied, namely,

$$x_1 = x_n, \quad x_1' = x_n', \quad x_1'' = x_n'', \qquad y_1 = y_n, \quad y_1' = y_n', \quad y_1'' = y_n'' \tag{8.2.1}$$

where x_i' and x_i'' are defined as

$$x_i' \equiv \left.\frac{dx}{dp}\right|_{x=x_i}, \qquad x_i'' \equiv \left.\frac{d^2x}{dp^2}\right|_{x=x_i}$$

with similar definitions for y_i' and y_i''. Conditions (8.2.1) can be satisfied by resorting to a periodic parametric cubic spline, as discussed in (Späth, 1978) and outlined below.

Let $(x,\ y)$ be the coordinates of one point of a cam profile. The curve describing the cam profile can be represented in parametric form as

$$x(p) = A_{xi}\,(p - p_i)^3 + B_{xi}\,(p - p_i)^2 + C_{xi}\,(p - p_i) + D_{xi} \tag{8.2.2a}$$

$$y(p) = A_{yi}\,(p - p_i)^3 + B_{yi}\,(p - p_i)^2 + C_{yi}\,(p - p_i) + D_{yi} \tag{8.2.2b}$$

$$\text{for} \quad p_i \le p \le p_{i+1}, \quad i = 1, \ldots, n-1$$

with the further definitions:

$$p_1 = 0, \quad p_{i+1} = p_i + \Delta p_i \tag{8.2.2c}$$

$$\Delta p_i = \sqrt{\Delta x_i^2 + \Delta y_i^2} \tag{8.2.2d}$$

$$\Delta x_i = x_{i+1} - x_i, \qquad \Delta y_i = y_{i+1} - y_i \tag{8.2.2e}$$

In the foregoing description, x_i and y_i represent the Cartesian coordinates of the ith supporting point (SP) of the spline, while coefficients A_{ui}, B_{ui}, C_{ui}, D_{ui}, for $u = x,\ y$, and $i = 1, \ldots, n-1$, are determined as explained presently. Let us define the $n'(\equiv n-1)$-dimensional vectors

$$\mathbf{x} \equiv \left[x_1, \ldots, x_{n'}\right]^T, \qquad \mathbf{y} \equiv \left[y_1, \ldots, y_{n'}\right]^T \tag{8.2.3a}$$

and

$$\mathbf{x}'' \equiv \left[x_1'', \ldots, x_{n'}''\right]^T, \qquad \mathbf{y}'' \equiv \left[y_1'', \ldots, y_{n'}''\right]^T \tag{8.2.3b}$$

The relationships between $\mathbf{x}$ and $\mathbf{x}''$ are linear (Späth, 1978), namely,

$$\mathbf{A}\,\mathbf{x}'' = 6\,\mathbf{C}\,\mathbf{x}, \qquad \mathbf{A}\,\mathbf{y}'' = 6\,\mathbf{C}\,\mathbf{y} \tag{8.2.4a}$$

which are expressions similar to those of eq. (2.4.3*b*), except that now the $\mathbf{A}$ and $\mathbf{C}$ matrices appearing in eq. (8.2.4*a*) are themselves functions of the coordinates of the SPs. In fact, the $n' \times n'$ matrices $\mathbf{A}$ and $\mathbf{C}$ are now defined as

$$\mathbf{A} = \begin{bmatrix} 2\alpha_{1,n'} & \alpha_1 & 0 & 0 & \cdots & \alpha_{n'} \\ \alpha_1 & 2\alpha_{1,2} & \alpha_2 & 0 & \cdots & 0 \\ 0 & \alpha_2 & 2\alpha_{2,3} & \alpha_3 & \cdots & 0 \\ \vdots & \vdots & \ddots & \ddots & \ddots & \vdots \\ 0 & 0 & \cdots & \alpha_{n'''} & 2\alpha_{n''',n''} & \alpha_{n''} \\ \alpha_{n'} & 0 & 0 & \cdots & \alpha_{n''} & 2\alpha_{n'',n'} \end{bmatrix} \tag{8.2.4b}$$

and

$$\mathbf{C} = \begin{bmatrix} -\beta_{1,n'} & \beta_1 & 0 & 0 & \cdots & \beta_{n'} \\ \beta_1 & -\beta_{1,2} & \beta_2 & 0 & \cdots & 0 \\ 0 & \beta_2 & -\beta_{2,3} & \beta_3 & \cdots & 0 \\ \vdots & \vdots & \ddots & \ddots & \ddots & \vdots \\ 0 & 0 & \cdots & \beta_{n'''} & -\beta_{n''',n''} & \beta_{n''} \\ \beta_{n'} & 0 & 0 & \cdots & \beta_{n''} & -\beta_{n'',n'} \end{bmatrix} \tag{8.2.4c}$$

where

$$\alpha_k = \Delta p_k, \quad \alpha_{i,j} = \alpha_i + \alpha_j, \quad \beta_k = 1/\alpha_k, \quad \beta_{i,j} = \beta_i + \beta_j; \qquad \text{for} \quad i, j, k = 1, \ldots, n' \tag{8.2.4d}$$

and

$$n' = n - 1, \qquad n'' = n - 2, \qquad n''' = n - 3 \tag{8.2.4e}$$

For the problem at hand, that is, the synthesis of the cam profile, we know the SPs, and hence, it is an easy task to obtain the coefficients for generating the closed curve. In fact, once the SPs are known, the coefficients A_{ui}, B_{ui}, C_{ui}, D_{ui}, for $u = x$, y, and $i = 1, \ldots, n-1$, can be generated as explained in Section 2.4 for non-parametric cubic splines, and can then produce the interpolated points. These points are meant for the actual display of the cam profile, and the generation of the cutter path of a machine tool if the follower is of the flat-face type. If the follower is of the roller type, the interpolated points represent the pitch curve, and a further step must be performed, namely, the addition of two vectors at each of the interpolated points, as outlined in Chapter 5.

8.3 General Transformation Formulas

The determination of the geometric properties of cams usually leads to problems associated with two- and three-dimensional regions. Thus, in this and the sections below, formulas are derived which allow the computation of the area—or the volume, in the case of 3-D regions—, centroid position vector, and inertia matrix of a closed region $\mathcal{R}$ bounded by surface $\mathcal{S}$.

Let $\mathcal{E}^{\nu}$ be a ν-dimensional Euclidean space, in which a bounded region $\mathcal{R}$ is embedded. The formulas presented here, whose derivation appears in (Angeles et al., 1990), are valid for Euclidean spaces of arbitrary (finite) dimension, our interest lying in the 2-D and 3-D cases, for which $\nu = 2$ and 3, respectively. A few definitions are first introduced.

The kth moment of $\mathcal{R}$ is defined as the integral given below:

$$\mathcal{I}_k = \int_{\mathcal{R}} f_k(\mathbf{r})\, d\mathcal{R}, \qquad k = 0, 1, \ldots \tag{8.3.1}$$

where $f_k(\mathbf{r})$ is a homogeneous function (Brand, 1965) of kth degree of the position vector $\mathbf{r}$. It is, moreover, a kth-rank *tensor*—i.e., every entry of this array contains k subscripts—, and hence $\mathcal{I}_k$ also is a kth-rank tensor. The most familiar moments are the first three, i.e., those defined for $k = 0$, 1, and 2. The zeroth moment $\mathcal{I}_0$ is simply the volume of $\mathcal{R}$, the first moment $\mathcal{I}_1$ being the vector producing the position vector $\bar{\mathbf{r}}$ of the centroid of $\mathcal{R}$ as

$$\bar{\mathbf{r}} = \frac{\mathcal{I}_1}{\mathcal{I}_0} \tag{8.3.2}$$

The second moment of $\mathcal{R}$, $\mathcal{I}_2$, is the inertia matrix of $\mathcal{R}$. The balance of this section assumes familiarity with multilinear algebra. The reader lacking this familiarity need only accept the relations appearing in eqs. (8.3.4a–c).

Let $\mathbf{\Phi}_m(\mathbf{r})$ be an mth-rank tensor function of $\mathbf{r}$ that is homogeneous of degree m in $\mathbf{r}$, $\operatorname{div}\mathbf{\Phi}_m$ denoting its divergence. Furthermore, let $(\cdot)$ denote the inner product of the tensor quantities beside it. The GDT states the relationship given below, namely,

$$\int_{\mathcal{R}} \operatorname{div}\mathbf{\Phi}_m\, d\mathcal{R} = \int_{\mathcal{S}} \mathbf{\Phi}_m \cdot \mathbf{n}\, d\mathcal{S} \tag{8.3.3}$$

where $\mathcal{S}$ and $\mathbf{n}$ denote the boundary of $\mathcal{R}$ and the outward unit normal of this boundary, respectively. If the function $f_k(\mathbf{r})$, whose integral is to be computed, is the *divergence* of function $\mathbf{\Phi}_m(\mathbf{r})$, then the said integral reduces to an integral on $\mathcal{S}$, by application of the GDT where, clearly, $m = k + 1$. However, finding a function $\mathbf{\Phi}_{k+1}(\mathbf{r})$ whose divergence be a given function $f_k(\mathbf{r})$ can be, in general, a more difficult task than computing the volume integral of $f_k(\mathbf{r})$ directly. Nevertheless, the computation of the moments of regions, particularly the first three moments, involves the derivation of $\mathbf{\Phi}_m(\mathbf{r})$ functions that can be readily obtained as described below. In fact, let $V_\nu, \mathbf{Q}_\nu^O$ and $\mathbf{I}_\nu^O$ denote the volume, the vector first moment, and the matrix second moment of $\mathcal{R}$, the last two of which are taken with respect to a given point O. The computation of these quantities can be reduced to that of integrals on $\mathcal{S}$ by application of the GDT, namely (Angeles et al., 1990),

$$V_\nu = \frac{1}{\nu} \int_{\mathcal{S}} \mathbf{r} \cdot \mathbf{n} \, d\mathcal{S} \tag{8.3.4a}$$

$$\mathbf{q}_\nu^O = \frac{1}{2} \int_{\mathcal{S}} (\mathbf{r} \cdot \mathbf{r}) \, \mathbf{n} \, d\mathcal{S} = \frac{1}{1+\nu} \int_{\mathcal{S}} \mathbf{r} \, (\mathbf{r} \cdot \mathbf{n}) \, d\mathcal{S} \tag{8.3.4b}$$

$$\mathbf{I}_\nu^O = \int_{\mathcal{S}} \mathbf{r} \cdot \mathbf{r} \left[\frac{3}{2\,(\nu+2)} \mathbf{1} \, (\mathbf{r} \cdot \mathbf{n}) - \frac{1}{2} \mathbf{r} \, \mathbf{n}^T \right] d\mathcal{S} \tag{8.3.4c}$$

where $\mathbf{1}$ denotes the identity matrix. In the sections below, the foregoing general formulas are applied to two- and three-dimensional regions whose boundaries are approximated using cubic spline functions. It is pointed out that $\mathbf{q}_\nu^O$ can be computed with two alternative formulas, which are given in eq. (8.3.4b). Moreover, the first formula is dimension-invariant, whereas the second one is more suitable for applications involving piecewise linear approximations, given the simple forms that the $\mathbf{r} \cdot \mathbf{n}$ term produces in such cases. Both formulas will prove to be useful in deriving simple practical formulas.

8.4 Two-Dimensional Regions

In this section the general formulas introduced above are applied to two-dimensional regions, i.e., $\nu = 2$. Thus, formulas will be derived for the computation of the area, the centroid position vector, and the inertia matrix of planar regions $\mathcal{R}$ bounded by a close contour Γ.

Let A, $\mathbf{q}^O$, and $\mathbf{I}^O$ denote the area, the vector first moment, and the matrix second moment of $\mathcal{R}$, $\mathbf{q}^O$ and $\mathbf{I}^O$ being defined with respect to a given point O in the plane of the region. The computation of these quantities can be reduced to integration on the boundary Γ by application of the GDT as outlined in (Angeles et al., 1990), namely,

$$A = \frac{1}{2} \int_\Gamma \mathbf{r} \cdot \mathbf{n} \, d\Gamma \tag{8.4.1}$$

$$\mathbf{q}^O = \frac{1}{2} \int_\Gamma (\mathbf{r} \cdot \mathbf{r}) \, \mathbf{n} \, d\Gamma = \frac{1}{3} \int_\Gamma \mathbf{r} \, (\mathbf{r} \cdot \mathbf{n}) \, d\Gamma \tag{8.4.2}$$

$$\mathbf{I}^O = \int_\Gamma \mathbf{r} \cdot \mathbf{r} \left[\frac{3}{8} \mathbf{1} \, (\mathbf{r} \cdot \mathbf{n}) - \frac{1}{2} \mathbf{r} \, \mathbf{n}^T \right] d\Gamma \tag{8.4.3}$$

where $\mathbf{1}$ denotes the 2×2 identity matrix and $\mathbf{n}$ is the unit outward normal of Γ, the products $\mathbf{r} \cdot \mathbf{n}$ and $\mathbf{r}\,\mathbf{n}^T$ being the inner product and the "tensor" product of vectors $\mathbf{r}$ and $\mathbf{n}$, already introduced in Section 3.3. These are reproduced below for quick reference, r_i, n_i, for $i = 1, 2$, denoting the components of vectors $\mathbf{r}$ and $\mathbf{n}$, correspondingly, in the given reference frame. Then,

$$\mathbf{r} \cdot \mathbf{n} = r_1 n_1 + r_2 n_2$$

and

$$\mathbf{r}\,\mathbf{n}^T = \begin{bmatrix} r_1 \\ r_2 \end{bmatrix} [n_1,\ n_2] = \begin{bmatrix} r_1 n_1 & r_1 n_2 \\ r_2 n_1 & r_2 n_2 \end{bmatrix}$$

In Subsections 8.4.1 and 8.4.2, the foregoing formulas, eqs. (8.4.1–3), are applied to planar regions with piecewise-linear and spline approximations of their boundaries, respectively. It is pointed out that $\mathbf{q}^O$ can be computed with two alternative formulas, which are given in eq. (8.4.2). We recall here that the second of these formulas is more suitable for applications involving piecewise linear approximations, given the simple forms that the $\mathbf{r} \cdot \mathbf{n}$ term produces in such cases. Both formulas will prove to be useful in deriving practical simple formulas, as shown below.

8.4.1 Piecewise-Linear Approximation of the Boundary

If Γ in eqs. (8.4.1)–(8.4.6) is approximated by a closed n-sided polygon, then

$$\Gamma \approx \bigcup_1^n \Gamma_i \tag{8.4.4}$$

where Γ_i denotes the ith side of the polygon. The aforementioned formulas, thus, can be approximated as

$$A \approx \frac{1}{2} \sum_1^n \int_{\Gamma_i} \mathbf{r} \cdot \mathbf{n}_i \, d\Gamma_i \tag{8.4.5a}$$

$$\mathbf{q}^O \approx \frac{1}{2} \sum_1^n \int_{\Gamma_i} (\mathbf{r} \cdot \mathbf{r})\, \mathbf{n}_i \, d\Gamma_i = \frac{1}{3} \sum_1^n \int_{\Gamma_i} \mathbf{r}\,(\mathbf{r} \cdot \mathbf{n}_i)\, d\Gamma_i \tag{8.4.5b}$$

$$\mathbf{I}^O \approx \sum_1^n \int_{\Gamma_i} \mathbf{r} \cdot \mathbf{r} \left[\frac{3}{8} \mathbf{1}\,(\mathbf{r} \cdot \mathbf{n}_i) - \frac{1}{2} \mathbf{r}\,\mathbf{n}_i^T \right] d\Gamma_i \tag{8.4.5c}$$

where $\mathbf{n}_i$ denotes the outward normal unit vector of Γ_i, and hence, it is a constant along this side of the approximating polygon.

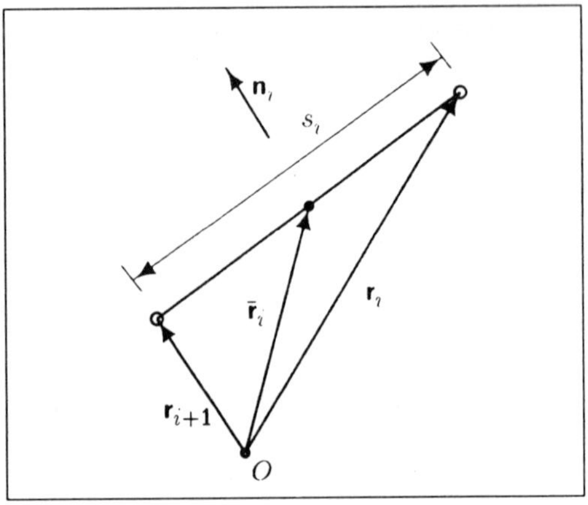

Fig. 8.4.1. Segment Γ_i of the boundary

Furthermore, let A_i, $\mathbf{q}_i^O$, and $\mathbf{I}_i^O$ be the contributions of Γ_i, the ith side of the polygon, to the corresponding integral, s_i and $\bar{\mathbf{r}}_i$ denoting its length and the position vector of its centroid, as shown in Fig. 8.4.1. From this figure, the reduced calculations are readily derived, namely,

$$A_i = \frac{1}{2}\mathbf{n}_i \cdot \int_{\Gamma_i} \mathbf{r}\, d\Gamma_i = \frac{1}{2}\mathbf{n}_i \cdot \bar{\mathbf{r}}_i\, s_i \tag{8.4.6}$$

Using each of the two formulas of eq. (8.4.5b), one obtains two alternate expressions for $\mathbf{q}_i^O$, namely,

$$\mathbf{q}_i^O = \frac{1}{2}\left(\int_{\Gamma_i} \mathbf{r}\cdot\mathbf{r}\, d\Gamma_i\right)\mathbf{n}_i \tag{8.4.7a}$$

$$\mathbf{q}_i^O = \frac{1}{3}\left(\int_{\Gamma_i} \mathbf{r}\,\mathbf{r}^T\, d\Gamma_i\right)\cdot\mathbf{n}_i \tag{8.4.7b}$$

By subtracting two times both sides of eq. (8.4.7a) from three times both sides of eq. (8.4.7b), one obtains

$$\mathbf{q}_i^O = -\mathbf{n}_i \cdot \int_{\Gamma_i} [(\mathbf{r}\cdot\mathbf{r})\mathbf{1} - \mathbf{r}\,\mathbf{r}^T]\, d\Gamma_i \tag{8.4.8}$$

The right-hand side of eq. (8.4.8) is readily recognized as the projection onto $-\mathbf{n}_i$ of the matrix second moment of segment Γ_i, with respect to O, represented here as $\mathbf{J}_i^O$. Thus,

$$\mathbf{q}_i^O = -\mathbf{n}_i \cdot \mathbf{J}_i^O \tag{8.4.9}$$

Furthermore,

$$\mathbf{I}_i^O = \int_{\Gamma_i} \mathbf{r}\cdot\mathbf{r}\left[\frac{3}{8}(\mathbf{r}\cdot\mathbf{n}_i)\,\mathbf{1} - \frac{1}{2}\,\mathbf{r}\,\mathbf{n}_i^T\right] d\Gamma_i = \frac{3}{8}\,\mathbf{1}\,(\mathbf{n}_i\cdot\mathbf{v}_i) - \frac{1}{2}\,\mathbf{v}_i\,\mathbf{n}_i^T \tag{8.4.10a}$$

with $\mathbf{v}_i$ defined as

$$\mathbf{v}_i = \int_{\Gamma_i} (\mathbf{r}\cdot\mathbf{r})\,\mathbf{r}\,d\Gamma_i \tag{8.4.10b}$$

Now, vector $\mathbf{r}$ appearing in the integrand of eq. (8.4.10b) is expressed as

$$\mathbf{r} = \mathbf{r}_i + \mu\,(\mathbf{r}_{i+1} - \mathbf{r}_i), \qquad 0 \le \mu \le 1 \tag{8.4.10c}$$

where $\mathbf{r}_i$ and $\mathbf{r}_{i+1}$ denote the position vectors of the end points of Γ_i. Thus, the foregoing vectors are the position vectors of the ith and the $(i+1)$st vertices of the approximating polygon, which are assumed to be numbered in counterclockwise order. Moreover, since the polygon is closed, $i+1 = 1$ for $i = n$ in eq. (8.4.10c) and below. Substitution of eq. (8.4.10c) into eq. (8.4.10b) yields, for $i = 1, \ldots, n$,

$$\mathbf{v}_i = \frac{s_i}{3}\,(A_i\,\mathbf{r}_i + B_i\,\mathbf{r}_{i+1}) \tag{8.4.10d}$$

where A_i and B_i are the scalars defined as

$$A_i \equiv \mathbf{r}_i\cdot\mathbf{r}_{i+1} + \frac{1}{2}\,\mathbf{r}_i\cdot\mathbf{r}_i + \frac{s_i^2}{4} \tag{8.4.10e}$$

$$B_i \equiv 2\,\mathbf{r}_i\cdot\mathbf{r}_{i+1} - \frac{1}{2}\,\mathbf{r}_i\cdot\mathbf{r}_i + \frac{3}{4}\,s_i^2 \tag{8.4.10f}$$

Formulas (8.4.6), (8.4.9), and (8.4.10a–f) are the relations sought.

8.4.2 Spline Approximation of the Boundary

The formulas introduced in Section 8.3 are now applied to regions whose boundary is approximated by periodic parametric cubic splines. Recalling expressions (8.2.1a & b), the $(x,\ y)$ coordinates of one point on the boundary are represented as functions of parameter p, while x_i and y_i represent the Cartesian coordinates of the ith supporting point of the spline. Moreover, integrals (8.4.1–3) are approximated as

$$A \approx \sum_1^{n-1} A_i \tag{8.4.11a}$$

$$Q_x^O \approx \sum_1^{n-1} Q_{xi}^O, \qquad Q_y^O \approx \sum_1^{n-1} Q_{yi}^O \tag{8.4.11b}$$

$$I_x^O \approx \sum_1^{n-1} I_{xi}, \quad I_{xy}^O \approx \frac{1}{2}\left(\sum_1^{n-1} I_{xyi} + \sum_1^{n-1} I_{yxi}\right), \quad I_y^O \approx \sum_1^{n-1} I_{yi} \tag{8.4.11c}$$

with

$$A_i = \frac{1}{2}\int_{p_i}^{p_{i+1}} (xy' - x'y)\, dp = \sum_{k=1}^{5} a_{ik}\,(\Delta p_i)^k \tag{8.4.12a}$$

$$Q_{xi}^O = \frac{1}{2}\int_{p_i}^{p_{i+1}} (x^2 + y^2)\, y'\, dp = \sum_{k=1}^{9} q_{xik}\,(\Delta p_i)^k \tag{8.4.12b}$$

$$Q_{yi}^O = -\frac{1}{2}\int_{p_i}^{p_{i+1}} (x^2 + y^2)\, x'\, dp = \sum_{k=1}^{9} q_{yik}\,(\Delta p_i)^k \tag{8.4.12c}$$

where

$$x' \equiv \frac{dx}{dp}, \qquad y' \equiv \frac{dy}{dp}$$

The polynomial coefficients a_{ik}, q_{xik}, and q_{yik}, appearing in eqs. (8.4.12*a–c*) are included in Appendix A. The components I_{xi}, I_{yi}, I_{xyi}, and I_{yxi} of the inertia matrix are computed as shown below:

$$\begin{aligned} I_{xi} &= \frac{1}{8}\int_{p_i}^{p_{i+1}} (x^2 + y^2)\,(x\, y' + 3x' y)\, dp \\ &= \frac{1}{8}\int_{p_i}^{p_{i+1}} (x^2 + y^2)\, x\, y'\, dp + \frac{3}{8}\int_{p_i}^{p_{i+1}} (x^2 + y^2)\, x'\, y\, dp \end{aligned} \tag{8.4.12d}$$

$$\begin{aligned} I_{yi} &= -\frac{1}{8}\int_{p_i}^{p_{i+1}} (x^2 + y^2)\,(3\, x\, y' + x' y)\, dp \\ &= -\frac{3}{8}\int_{p_i}^{p_{i+1}} (x^2 + y^2)\, x\, y'\, dp - \frac{1}{8}\int_{p_i}^{p_{i+1}} (x^2 + y^2)\, x'\, y\, dp \end{aligned} \tag{8.4.12e}$$

In the second of formulas (8.4.11*c*), one summation would suffice to compute the product of inertia I_{xy}^O, given the symmetry of the inertia matrix. The two summations are computed here, and then averaged, in order to diminish roundoff or measurement errors that could be introduced if, for instance, the coordinates of the cam profile—the supporting points—were digitized. Thus, the terms of both summations are given by

$$I_{xyi} = \frac{1}{2}\int_{p_i}^{p_{i+1}} (x^2 + y^2)\, x\, x'\, dp \tag{8.4.12f}$$

$$I_{yxi} = -\frac{1}{2}\int_{p_i}^{p_{i+1}} (x^2 + y^2)\, y\, y'\, dp \tag{8.4.12g}$$

From expressions (8.4.12d–g) it is noted that all components of the inertia matrix are given as linear combinations of four integrals, namely,

$$I_i^{(1)} \equiv \int_{p_i}^{p_{i+1}} (x^2 + y^2)\, x\, x'\, dp = \sum_1^{12} h_{ik}^{(1)} (\Delta p_i)^k \tag{8.4.12h}$$

$$I_i^{(2)} \equiv \int_{p_i}^{p_{i+1}} (x^2 + y^2)\, y\, y'\, dp = \sum_1^{12} h_{ik}^{(2)} (\Delta p_i)^k \tag{8.4.12i}$$

$$I_i^{(3)} \equiv \int_{p_i}^{p_{i+1}} (x^2 + y^2)\, x'\, y\, dp = \sum_1^{12} h_{ik}^{(3)} (\Delta p_i)^k \tag{8.4.12j}$$

$$I_i^{(4)} \equiv \int_{p_i}^{p_{i+1}} (x^2 + y^2)\, x\, y'\, dp = \sum_1^{12} h_{ik}^{(4)} (\Delta p_i)^k \tag{8.4.12k}$$

The polynomial coefficients $h_{ik}^{(1)}, \ldots, h_{ik}^{(4)}$ appearing in eqs. (8.4.12h–k) are included in Appendix A.

8.5 Three-Dimensional Regions

For solid regions $\nu = 3$, the general relations (8.3.4a–c) thus are reduced to:

$$V_3 = \frac{1}{3}\int_S \mathbf{r}\cdot\mathbf{n}\, dS \tag{8.5.1a}$$

$$\mathbf{q}_3^O = \frac{1}{2}\int_S (\mathbf{r}\cdot\mathbf{r})\,\mathbf{n}\, dS = \frac{1}{4}\int_S \mathbf{r}\,(\mathbf{r}\cdot\mathbf{n})\, dS \tag{8.5.1b}$$

$$\mathbf{I}_3^O = \int_S \mathbf{r}\cdot\mathbf{r}\left[\frac{3}{10}\mathbf{1}\,(\mathbf{r}\cdot\mathbf{n}) - \frac{1}{2}\mathbf{r}\,\mathbf{n}^T\right] dS \tag{8.5.1c}$$

Next, explicit formulas are presented that are applicable to piecewise-linear approximations of boundaries of solids of arbitrary shapes.

One simple approximation of the boundary can be obtained by means of a polyhedron formed by polygonal faces. The integrals appearing in eqs. (8.5.1a–c) therefore can be expressed as sums of integrals over the polyhedral faces, the whole boundary $\mathcal{S}$ thus being approximated as

$$\mathcal{S} \approx \bigcup_1^n \mathcal{S}_i \tag{8.5.2}$$

where each part $\mathcal{S}_i$ is a polygonal portion of a plane. The integral formulas, thus, can be approximated as

$$V_3 \approx \frac{1}{3}\sum_1^n \int_{\mathcal{S}_i} \mathbf{r}\cdot\mathbf{n}_i \, d\mathcal{S}_i \tag{8.5.3a}$$

$$\mathbf{q}_3^O \approx \frac{1}{2}\sum_1^n \int_{\mathcal{S}_i} (\mathbf{r}\cdot\mathbf{r})\,\mathbf{n}_i \, d\mathcal{S}_i = \frac{1}{4}\sum_1^n \int_{\mathcal{S}_i} \mathbf{r}\,(\mathbf{r}\cdot\mathbf{n}_i)\, d\mathcal{S}_i \tag{8.5.3b}$$

$$\mathbf{I}_3^O \approx \sum_1^n \int_{\mathcal{S}_i} \mathbf{r}\cdot\mathbf{r}\left[\frac{3}{10}\,\mathbf{1}\,(\mathbf{r}\cdot\mathbf{n}_i) - \frac{1}{2}\mathbf{r}\,\mathbf{n}_i^T\right] d\mathcal{S}_i \tag{8.5.3c}$$

Now, let V_{3i}, $\mathbf{q}_{3i}^O$ and $\mathbf{I}_{3i}^O$ be the contribution of ith face, $\mathcal{S}_i$, of the polyhedron to the corresponding integral, Δ_i, $\bar{\mathbf{r}}_i$, and $\mathbf{I}_{\Delta_i}^O$ being the area, the position vector of the centroid, and the inertia matrix of the polygon $\mathcal{S}_i$, respectively, the last two quantities being taken with respect to O. Then,

$$V_{3i} = \frac{1}{3}\mathbf{n}_i \cdot \int_{\mathcal{S}_i} \mathbf{r}\, d\mathcal{S}_i = \frac{1}{3}\mathbf{n}_i \cdot \bar{\mathbf{r}}_i\, \Delta_i \tag{8.5.4}$$

We calculate the polygon area Δ_i and its centroid $\bar{\mathbf{r}}_i$ in a plane defined by the polygon, using the method applied to 2-D regions as outlined in Section 8.4.

By subtracting twice the rightmost-side of (8.5.3b) from the middle one, we readily obtain an expression for $\mathbf{q}_3^O$ as a summation of terms $\mathbf{q}_{3i}^O$ having the form:

$$\begin{aligned} \mathbf{q}_{3i}^O &= -\frac{1}{2}\int_{\mathcal{S}_i} [\,(\mathbf{r}\cdot\mathbf{r})\,\mathbf{n}_i - (\mathbf{r}\cdot\mathbf{n}_i)\,\mathbf{r}\,]\, d\mathcal{S}_i \\ &= -\frac{1}{2}\mathbf{n}_i \cdot \int_{\mathcal{S}_i} [\,(\mathbf{r}\cdot\mathbf{r})\,\mathbf{1} - \mathbf{r}\,\mathbf{r}^T\,]\, d\mathcal{S}_i \end{aligned} \tag{8.5.5}$$

The second integral of eq. (8.5.5) is readily identified as $\mathbf{I}_{\Delta_i}^O$—the second moment of polygon $\mathcal{S}_i$. Hence,

$$\mathbf{q}_{3i}^O = -\frac{1}{2}\,\mathbf{n}_i \cdot \mathbf{I}_{\Delta i}^O \tag{8.5.6}$$

i.e., the contribution of $\mathcal{S}_i$ to the first moment of $\mathcal{R}$ is recognized as the projection onto $-\frac{1}{2}\,\mathbf{n}_i$ of the second moment of $\mathcal{S}_i$, both moments being taken, of course, with respect to the same point O. We calculate the centroidal inertia matrix of the polygon at hand using formulas derived for 2-D regions in a plane defined by the polygon. Using the parallel-axes theorem and a rotation of axes, we then find $\mathbf{I}_{\Delta_i}^O$ from the calculated centroidal inertia matrix.

Additionally, the contribution of the ith polygonal face to the second moment of the 3-D region under study is derived below:

$$\mathbf{I}_{3i}^O = \frac{1}{2}\int_{\mathcal{S}_i} \mathbf{r}\cdot\mathbf{r}\left[\frac{3}{5}(\mathbf{r}\cdot\mathbf{n}_i)\,\mathbf{1} - \mathbf{r}\,\mathbf{n}_i^T\right] d\mathcal{S}_i$$

$$= \frac{3}{10}(\mathbf{n}_i\cdot\mathbf{w}_i)\,\mathbf{1} - \frac{1}{2}\,\mathbf{w}_i\,\mathbf{n}_i^T \tag{8.5.7a}$$

with $\mathbf{w}_i$ defined as:

$$\mathbf{w}_i \equiv \int_{\mathcal{S}_i} (\mathbf{r}\cdot\mathbf{r})\,\mathbf{r}\,d\mathcal{S}_i \tag{8.5.7b}$$

The integral appearing in eq.(8.5.7b) is evaluated below. To this end, $\mathbf{r}$ is expressed as:

$$\mathbf{r} = \overline{\mathbf{r}}_i + \mathbf{p} \tag{8.5.8}$$

where $\mathbf{p}$ is a vector lying in the plane of the polygon $\mathcal{S}_i$ and stemming from the polygon centroid, as shown in Fig. 8.5.1. Now, $\mathbf{w}_i$ becomes

$$\mathbf{w}_i = \overline{\mathbf{r}}_i^2\,\Delta_i\overline{\mathbf{r}} + \overline{\mathbf{r}}_i\cdot\int_{\mathcal{S}_i}(2\,\mathbf{p}\,\mathbf{p}^T + \mathbf{p}^2\mathbf{1})\,d\mathcal{S}_i + \int_{\mathcal{S}_i}\mathbf{p}^2\mathbf{p}\,d\mathcal{S}_i \tag{8.5.9}$$

where an exponent k over a vector quantity indicates the kth power of the magnitude of the said vector, i.e., $\mathbf{r}^2 \equiv \mathbf{r}\cdot\mathbf{r} = \|\mathbf{r}\|^2$, a notation that will be used in the discussion below.

Three surface integrals over $\mathcal{S}_i$ need to be evaluated in the expression for $\mathbf{w}_i$, namely,

$$\mathbf{A}_i = \int_{\mathcal{S}_i}\mathbf{p}\,\mathbf{p}^T\,d\mathcal{S}_i \tag{8.5.10a}$$

$$a_i = \int_{\mathcal{S}_i}\mathbf{p}^2\,d\mathcal{S}_i \tag{8.5.10b}$$

$$\mathbf{b}_i = \int_{\mathcal{S}_i}\mathbf{p}^2\,\mathbf{p}\,d\mathcal{S}_i \tag{8.5.10c}$$

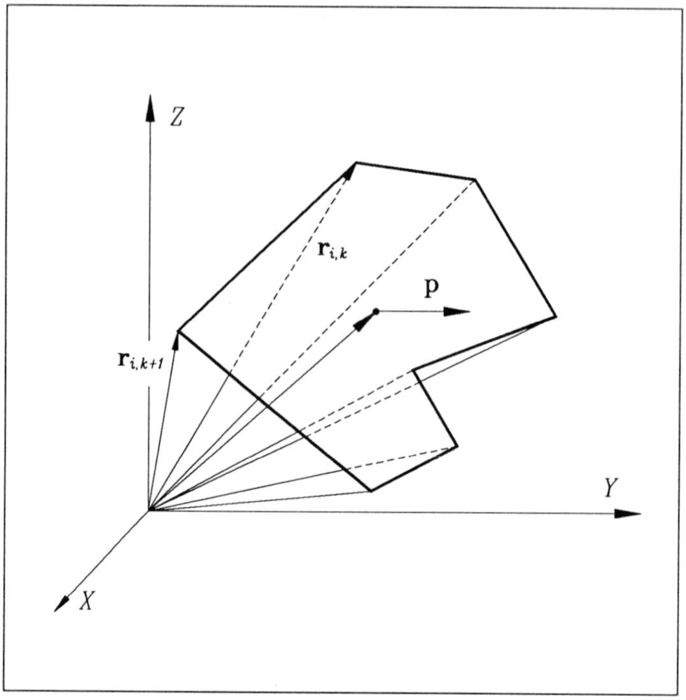

Fig. 8.5.1. Polygon representing one face of a polyhedron approximating a closed surface.

Since $\mathbf{p}$ is a vector lying entirely in the plane Π_i defined by the polygon $\mathcal{S}_i$, it can be represented uniquely in the 2-D subspace as a 2-D vector of that plane. Consequently, the GDT can be applied in this 2-D subspace to reduce the surface integrals (8.5.10a–c) to the following line integrals (Angeles and Al-Daccak, 1989):

$$\mathbf{A}_i = \frac{1}{4}\int_{\mathcal{S}_i} \operatorname{div}(\mathbf{P})\, d\mathcal{S}_i = \frac{1}{4}\int_{\Gamma_i} \mathbf{p}\,\mathbf{p}^T (\mathbf{p}\cdot\overline{\mathbf{n}}_i)\, d\Gamma_i \qquad (8.5.11a)$$

$$a_i = \frac{1}{4}\int_{\mathcal{S}_i} \operatorname{div}(\mathbf{p}^2\mathbf{p})\, d\mathcal{S}_i = \frac{1}{4}\int_{\Gamma_i} \mathbf{p}^2 (\mathbf{p}\cdot\overline{\mathbf{n}}_i)\, d\Gamma_i \qquad (8.5.11b)$$

$$\mathbf{b}_i = \frac{1}{5}\int_{\mathcal{S}_i} \operatorname{div}(\mathbf{p}^2\,\mathbf{p}\,\mathbf{p}^T)\, d\mathcal{S}_i = \frac{1}{5}\int_{\Gamma_i} \mathbf{p}^2\mathbf{p}\,(\mathbf{p}\cdot\overline{\mathbf{n}}_i)\, d\Gamma_i \qquad (8.5.11c)$$

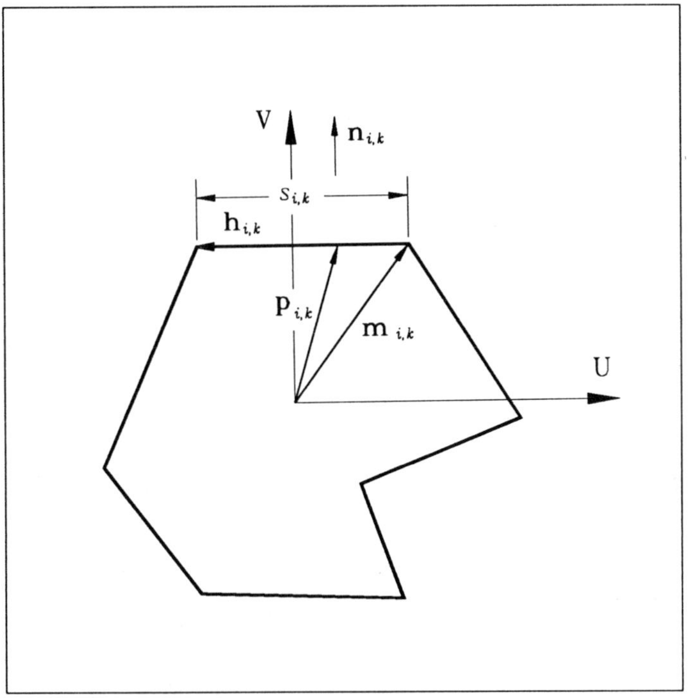

Fig. 8.5.2. The ith polygon contained in plane Π_i.

where $\mathbf{P}$ is a *third-rank* tensor that is cubic and homogeneous in $\mathbf{p}$. It is omitted here, the interested reader being referred to the above-mentioned reference for further details. Moreover, Γ_i denotes the polygonal boundary of $\mathcal{S}_i$ and $\overline{\mathbf{n}}_i$ the outward unit normal vector of Γ_i, contained in plane Π_i. Note that, just as the GDT is used in 3-D to reduce volume integrals to surface integrals, it is used in 2-D to reduce surface integrals to line integrals.

Now, let $\Gamma_{i,k}$ denote the kth side of the m-sided polygon $\mathcal{S}_i$ that joins the kth and the $(k+1)$st vertices, numbered counterclockwise when the face of interest is viewed from outside the polyhedron. Moreover, a sum over subscript k is to be understood, henceforth, as being *modulo* m. Furthermore, the position vector $\mathbf{p}$, shown in Fig. 8.5.2, of any point of $\Gamma_{i,k}$ is defined in plane Π_i as

$$\mathbf{p} = \mathbf{m}_k + \mathbf{h}_k\, s, \qquad \mathbf{p} \in \Gamma_{i,k}, \qquad 0 \leq s \leq 1 \tag{8.5.12}$$

where $\mathbf{m}_k$ and $\mathbf{h}_k$ are the vectors $(\mathbf{r}_{i,k} - \overline{\mathbf{r}}_i)$ and $(\mathbf{r}_{i,k+1} - \mathbf{r}_{i,k})$, respectively, $\mathbf{r}_{i,k}$ being the position vector of the kth vertex of polygon $\mathcal{S}_i$. Similar to vector $\mathbf{p}$, vectors $\mathbf{m}_k$ and $\mathbf{h}_k$ lie solely in the plane Π_i. Consequently, their representation as 2-D vectors in that plane is used in order to apply the GDT to reduce the surface integrals defined over $\mathcal{S}_i$ to line integrals defined over $\Gamma_{i,k}$.

Let $\mathbf{n}_{i,k}$ be the unit normal vector to $\Gamma_{i,k}$—pointing outwards of $\mathcal{S}_i$ —, and $s_{i,k}$ be the length of the kth side of Γ_i. Thus, quantities $\mathbf{A}_i$, a_i, and $\mathbf{b}_i$ can be evaluated as indicated below, keeping in mind that $s_{i,k} = \|\mathbf{h}_k\|$ and $\mathbf{p}\cdot\mathbf{n}_{i,k} = \mathbf{m}_k\cdot\mathbf{n}_{i,k}$, since $\mathbf{h}_k\cdot\mathbf{n}_{i,k} = 0$. Then,

$$\mathbf{A}_i = \frac{1}{4}\sum_{k=1}^{m} s_{i,k}\,(\mathbf{m}_k\cdot\mathbf{n}_{i,k})\int_0^1 \mathbf{p}\,\mathbf{p}^T\,ds, \quad \mathbf{p}\in\Gamma_{i,k} \tag{8.5.13a}$$

where

$$\mathbf{p}\,\mathbf{p}^T = (\mathbf{m}_k\,\mathbf{m}_k^T) + (\mathbf{m}_k\,\mathbf{h}_k^T + \mathbf{h}_k\,\mathbf{m}_k^T)\,s + (\mathbf{h}_k\,\mathbf{h}_k^T)\,s^2 \tag{8.5.13b}$$

and hence,

$$\int_0^1 \mathbf{p}\,\mathbf{p}^T\,ds = (\mathbf{m}_k\,\mathbf{m}_k^T) + \frac{1}{2}(\mathbf{m}_k\,\mathbf{h}_k^T + \mathbf{h}_k\,\mathbf{m}_k^T) + \frac{1}{3}(\mathbf{h}_k\,\mathbf{h}_k^T) \tag{8.5.13c}$$

Also,

$$a_i = \frac{1}{4}\sum_{k=1}^{m} s_{i,k}\,(\mathbf{m}_k\cdot\mathbf{n}_{i,k})\int_0^1 \mathbf{p}^2\,ds, \quad \mathbf{p}\in\Gamma_{i,k} \tag{8.5.14a}$$

where

$$\mathbf{p}^2 = \mathbf{m}_k^2 + 2\mathbf{m}_k\cdot\mathbf{h}_k\,s + \mathbf{h}_k^2\,s^2 \tag{8.5.14b}$$

and hence,

$$\int_0^1 \mathbf{p}^2\,ds = \mathbf{m}_k^2 + \mathbf{m}_k\cdot\mathbf{h}_k + \frac{1}{3}\mathbf{h}_k^2 \tag{8.5.14c}$$

Furthermore,

$$\mathbf{b}_i = \frac{1}{5}\sum_{k=1}^{m} s_{i,k}\,(\mathbf{m}_k\cdot\mathbf{n}_{i,k})\int_0^1 \mathbf{p}^2\mathbf{p}\,ds, \quad \mathbf{p}\in\Gamma_{i,k} \tag{8.5.15a}$$

where

$$\mathbf{p}^2\mathbf{p} = \mathbf{p}^2\mathbf{m}_k + \mathbf{p}^2\mathbf{h}_k\,s \tag{8.5.15b}$$

and hence,

$$\int_0^1 \mathbf{p}^2\mathbf{p}\,ds = \mathbf{m}_k\left(\mathbf{m}_k^2 + \mathbf{m}_k\cdot\mathbf{h}_k + \frac{1}{3}\mathbf{h}_k^2\right) + \\ + \mathbf{h}_k\left(\frac{1}{2}\mathbf{m}_k^2 + \frac{2}{3}\mathbf{m}_k\cdot\mathbf{h}_k + \frac{1}{4}\mathbf{h}_k^2\right) \tag{8.5.15c}$$

Now that the three surface integrals given in eqs. (8.5.10a–c) have been reduced to line integrals and evaluated in plane Π_i, the results obtained in this 2-D subspace should be mapped to produce the results necessary for calculating $\mathbf{w}_i$ in the 3-D space, eq. (8.5.9). The scalar quantity a_i poses no problems and it is readily multiplied by the 3×3 identity matrix $\mathbf{1}$ in the expression for $\mathbf{w}_i$. Furthermore, matrix $\mathbf{A}_i$ and vector $\mathbf{b}_i$ in the 2-D subspace are transformed to their matrix and vector counterparts, respectively, in the 3-D space, before being substituted in the expression for $\mathbf{w}_i$.

This completes the calculation of $\mathbf{w}_i$, in terms of which we can calculate the matrix second moment of the piecewise-linear approximation of $\mathcal{R}$—see eqs.(8.5.7a,b). The above formulas have been implemented in a C code (Al-Daccak, 1989) for computing the global volumetric properties of arbitrary solids.

8.6 Examples

In the examples given below, the global geometric properties of an actual cam are computed. That is, the hub and its keyway are included. Thus, the cam is an actual 3-D solid that cannot be generated by simple extrusion.

8.6.1 Example 8.6.1: Global Geometric Properties of a Cam Disk for a Translating Flat-Face Follower

Three views of the actual cam obtained from Example 4.2.2 are shown in Fig. 8.6.1. For that cam, determine its volume, centroid location, inertia matrix about the centroid, and principal axes and moments of inertia of this matrix.

Throughout we will use the notation indicated below and illustrated in Figs. 8.6.1 and 8.6.2:

a: keyway width
b: keyway depth
r_b: base radius
r_i: internal hub radius
r_o: external hub radius
t_1: cam thicknes
t_2: hub thicknes

Solution:

The database generated from Example 4.2.2 was input to MOMENT, along with the numerical values given below in mm:

$a = 5.86$
$b = 3.51$
$r_b = 28.00$

$r_i = 11.62$
$r_o = 24.61$
$t_1 = 17.58$
$t_2 = 17.58$

With the above data, a solid model was produced, and its global volumetric properties were computed. These are

Volume:	183231.6 mm^3
Centroid coordinates:	$\bar{x} = -8.647$ mm,
	$\bar{y} = -24.085$ mm,
	$\bar{z} = -6.399$ mm,
Centroidal principal axes of inertia:	E_1, E_2, and E_3 (See Fig. 8.6.2)
Inertia matrix about its centroid	

$$\mathbf{I}_C = \begin{bmatrix} 14.01 & -0.179 & 0.375 \\ -0.179 & 15.79 & 1.061 \\ 0.375 & 1.061 & 27.52 \end{bmatrix} \times 10^7 \text{mm}^4$$

Principal moments of inertia:	$I_1 = 1.398 \times 10^8 \text{ mm}^4$,
	$I_2 = 1.571 \times 10^8 \text{ mm}^4$
	$I_3 = 2.763 \times 10^8 \text{ mm}^4$

8.6.2 Example 8.6.2: Global Geometric Properties of a Cam Disk for an Oscillating Knife-Edge Follower

Three views of the normalized cam obtained from Example 5.3.1 are shown in Fig. 8.6.3. Determnine the global geometric properties as in Example'8.6.1, with the nondimensional numerical values given below:

$a = 0.10$
$b = 0.06$
$r_b = 0.48$
$r_i = 0.20$
$r_o = 0.42$
$t_1 = 0.30$
$t_2 = 0.30$

Solution:

The database generated from Example 5.3.1 was input to MOMENT along with the given numerical values. With the above data, a solid model was produced, and its global volumetric properties were computed. These are

Volume:	$0.694\ \text{mm}^3$
Centroid coordinates:	$\overline{x} = -0.229$ mm,
	$\overline{y} = -0.029$ mm,
	$\overline{z} = -0.095$ mm,
Centroidal principal axes of inertia:	E_1, E_2, and E_3 (See Fig. 8.6.4)

Inertia matrix about its centroid,

$$\mathbf{I}_C \begin{bmatrix} 14.25 & -0.774 & 0.884 \\ -0.774 & 10.52 & 0.110 \\ 0.884 & 0.110 & 21.86 \end{bmatrix} \times 10^{-2}\text{mm}^4$$

Principal moments of inertia:	$I_1 = 0.1036\ \text{mm}^4$,
	$I_2 = 0.1429\ \text{mm}^4$
	$I_3 = 0.2196\ \text{mm}^4$

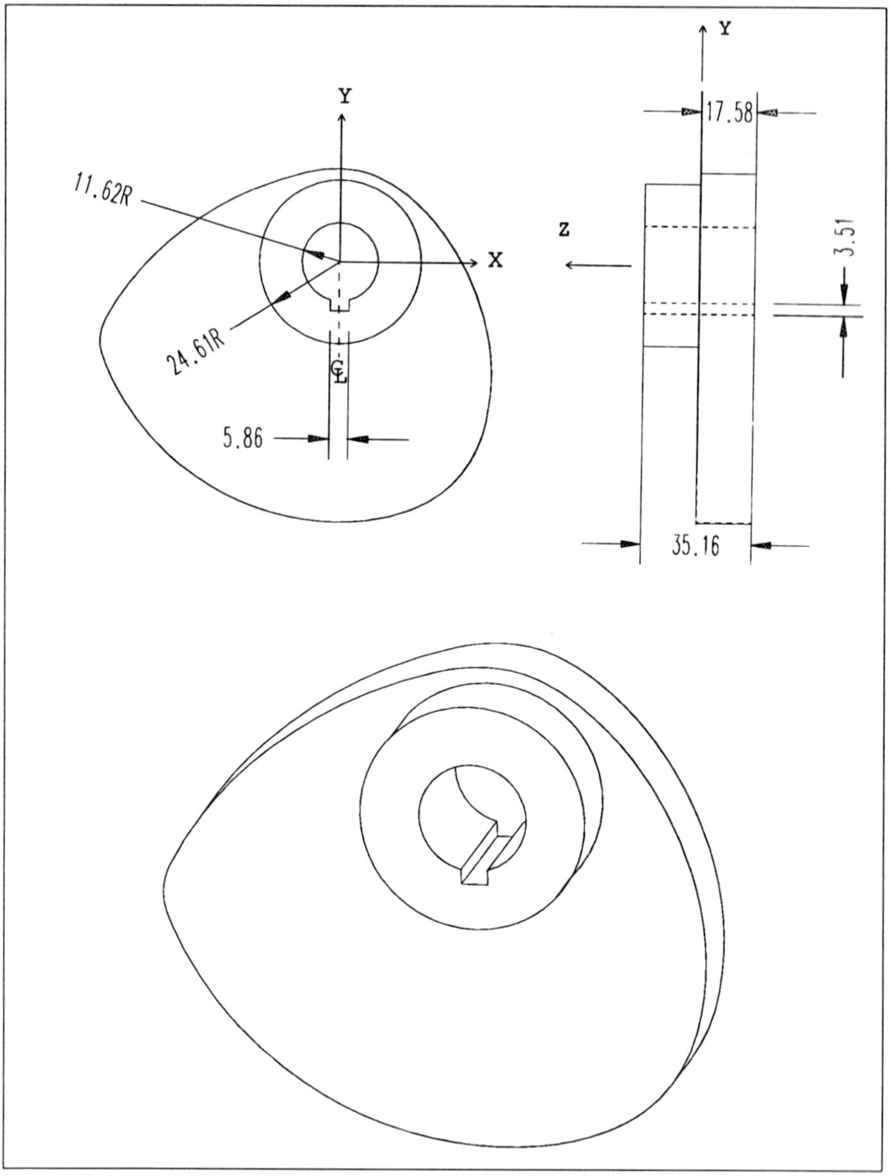

Fig. 8.6.1. Views of the cam for Example 8.6.1.

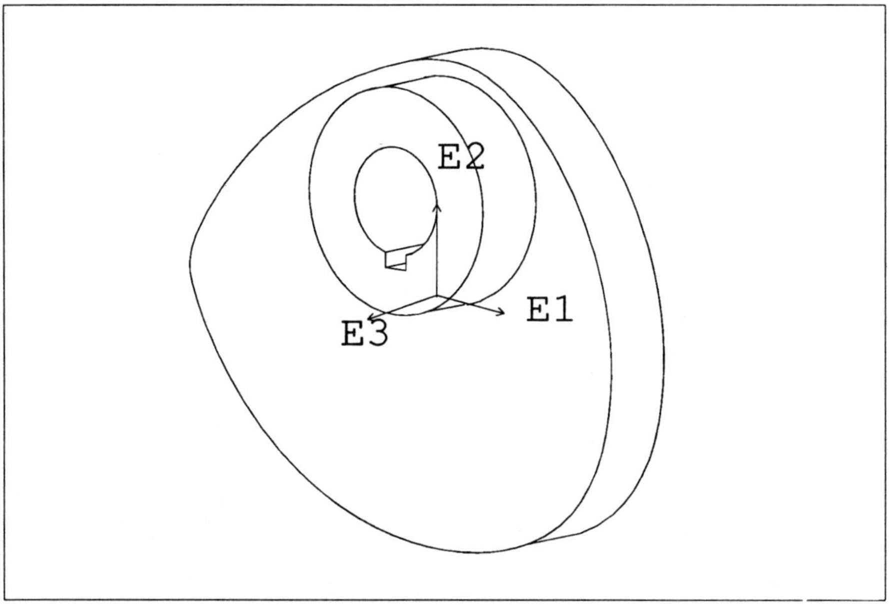

Fig. 8.6.2. Centroid and centroidal axes of inertia for Example 8.6.1.

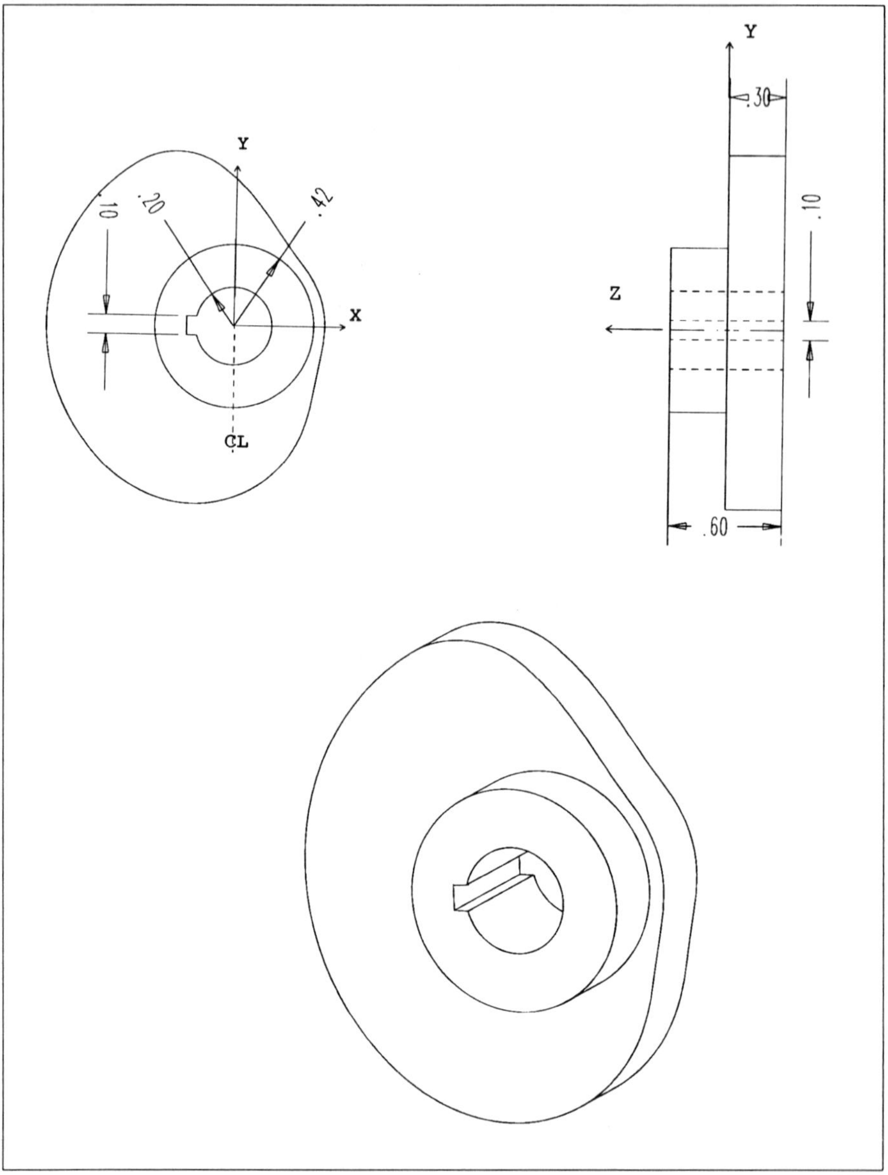

Fig. 8.6.3. Views of the cam for Example 8.6.2.

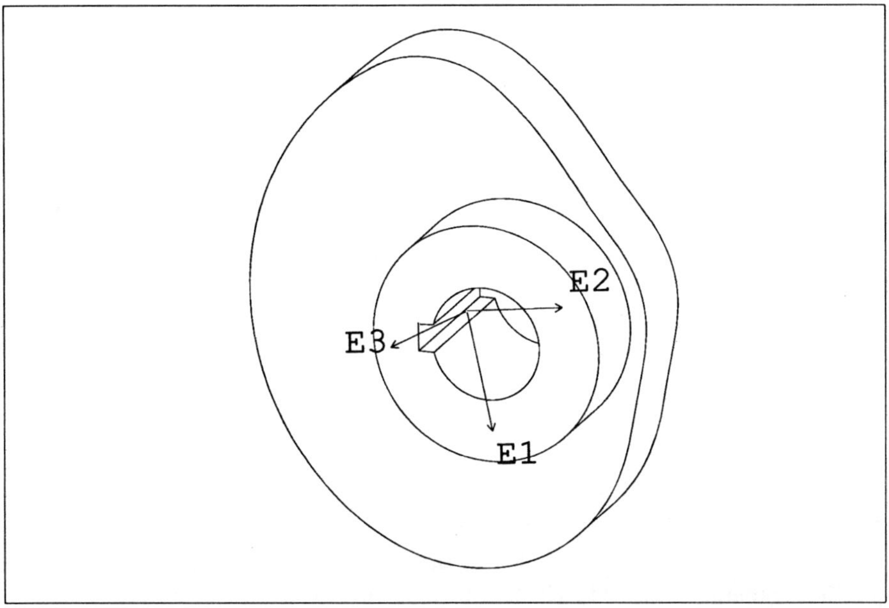

Fig. 8.6.4. Centroid and centroidal axes of inertia for Example 8.6.2.

Chapter 9 THE COMPUTER-AIDED DRAFTING AND MANUFACTURE OF CAMS

9.1 Introduction

The rendering of a cam profile, whether in the form of a blueprint or as a metal or plastic workpiece, has posed a challenging problem to engineers over the years. However, with the development of computer peripherals, such as the cathode-ray tube (CRT), the plotter, and the numerically-controlled (NC) machine tool, these tasks have become routine work. In fact, contrary to the most common machine elements, which involve either straight-line segments or circular arcs, cam profiles contain rather complex shapes which cannot be produced accurately with conventional—i.e., non-digital—drafting and machining facilities. This chapter focuses on the rendering of cams using a computer, but conventional methods are briefly reviewed, for completeness.

9.2 The Computer-Aided Rendering of Cam Profiles

The rendering of cams used to be a tremendously difficult and time-consuming task prior to the development of CAD/CAM techniques. This difficulty came about because cams usually contain not only simple *primitives* such as lines and circular arcs, or planes and cylinders in the 3-D case, but also *complex shapes.* By the latter we mean curves that do not have a simple parametric representation in the form of a polynomial or harmonic function.

In the past cam contours were produced with the sole aid of drafting instruments using geometrical construction techniques (Rothbahrt, 1956). These techniques are based on *kinematic inversion*, whereby the cam plate is held fixed while the machine frame is rotated in the direction opposite to that of the cam plate. In this way, the guideway of a translating follower, or the pivot of its oscillating counterpart, moves around the cam plate, the follower thus occupying a set of relative positions and orientations with respect to the cam plate. For translating flat-face followers, Fig. 9.2.1, the follower face is displayed in a discrete set of configurations L_1, L_2, ..., L_n. The cam contour is then produced with the aid of a French curve or a spline that traces a smooth curve tangent to the lines $\{L_k\}_1^n$.

For an oscillating knife-edge follower, the follower pivot is displayed in several positions P_1, P_2, ..., P_n around the center of rotation of the cam. Next, a line fixed to the follower arm and passing through the pivot P and the knife-edge K is displayed in its various configurations P_1K_1, P_2K_2, ..., P_nK_n. The set of points $\{K_k\}_1^n$ determines the cam profile, which is obtained by drawing a smooth curve passing through that set, as shown in Fig. 9.2.2. Alternatively, if the oscillating follower is coupled to the cam through a roller of radius a and center K, then the roller is displayed in its various configurations, as shown in Fig. 9.2.3. The cam profile is then produced by drawing a smooth curve tangent to all the displayed roller configurations.

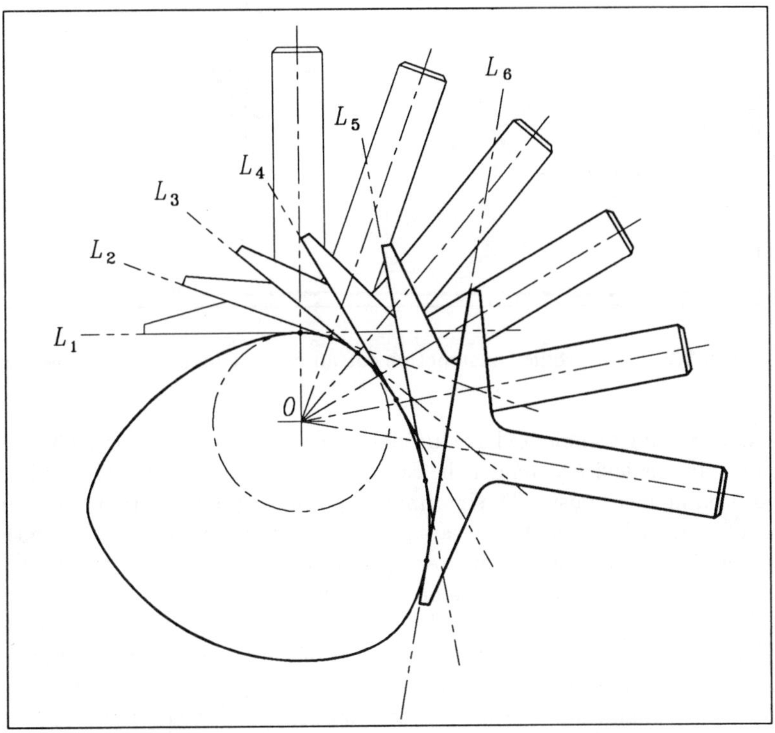

Fig. 9.2.1. Cam-profile construction for a translating flat-face follower.

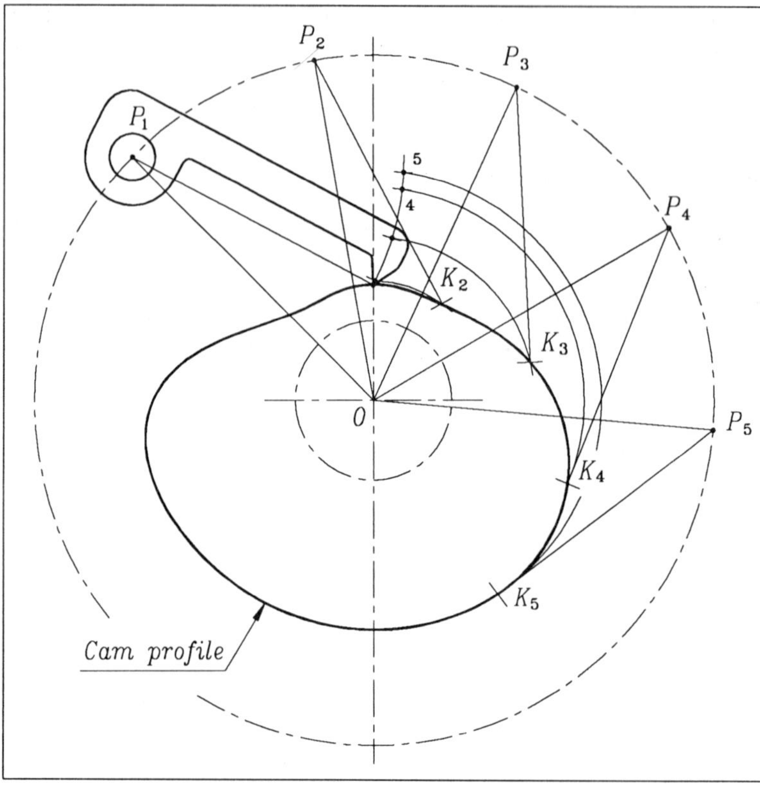

Fig. 9.2.2. Cam-profile construction for an oscillating knife-edge follower.

These drafting techniques have limited accuracy. On the contrary, when we use a computer-driven device to render the cam profile, accuracy is limited only by that of the positioning mechanism. The information pertaining to the various configurations of the follower with respect to the cam profile is supplied to the computer in the form of a database of Cartesian coordinates either of the cam profile or of a characteristic point of the cutting tool. Usually one employs a milling machine when cutting a cam plate, the tool being a circular disk of a suitable size. In this case, the database contains a sample of the points of the trajectory of the center of the mill. For purposes of display using a plotter, the database contains a sample of points of the cam profile itself and determines the trajectory of the tracing pen. When we display the profile on a monitor, the database determines the *pixel* location on the coordinate frame defined on the monitor. A pixel is a *picture element* in the terminology of computer graphics. It denotes the most primitive element in this context and is the equivalent of a point in planar geometry. Unlike a point, a pixel has physical dimensions. The larger the number of pixels in a monitor, the better the *resolution* of the device. Pixels on a monitor are arrayed on lines in the horizontal and the vertical directions, the resolution being usually indicated in the form $L_h \times L_v$ lines where L_h and L_v denote, respectively, the number of lines in the horizontal and vertical directions. Common values of L_h range from 640 to 1024, those of L_v from 200 to 1024.

We now present a brief description of the hardware required for a typical interactive graphics system, and a survey of the commercially available software for the displaying and plotting of cam profiles—and for any other similar application. We also give an outline of curve-smoothing techniques commonly used in computer graphics for the rendering of 3-D curves and surfaces.

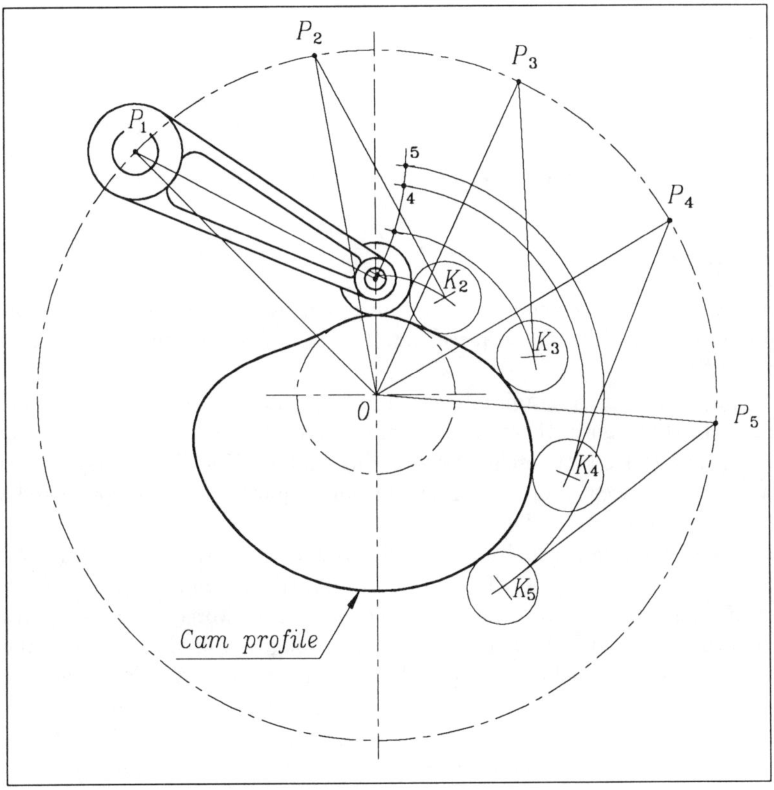

Fig. 9.2.3. Cam-profile construction for an oscillating roller-follower.

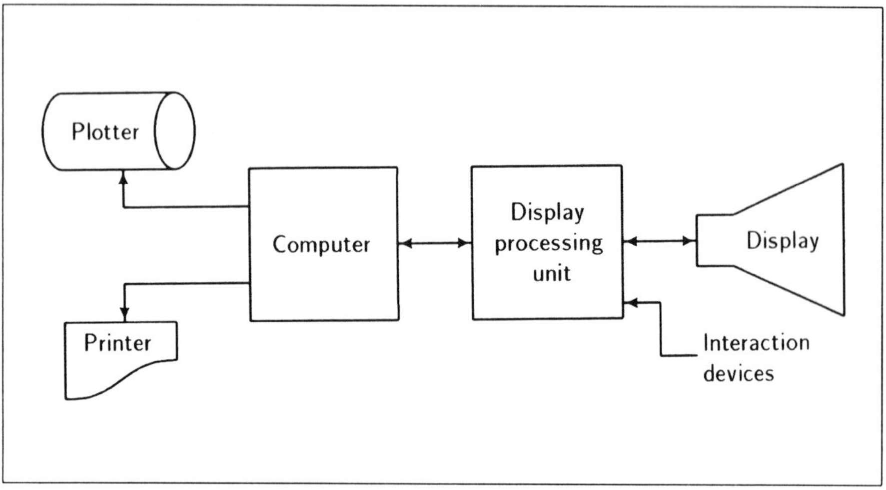

Fig. 9.2.4. A typical interactive graphics system.

9.2.1 Hardware

We give below a brief account of the graphics hardware available. More details are available in the specialized literature, e.g. (Foley and Van Dam, 1983; Salmon and Slater, 1987).

A typical interactive graphics system (Foley and Van Dam, 1983) is shown in Fig. 9.2.4. In that figure, there are four major subsystems, the computer or central processing unit (CPU), the display processing unit (DPU), the display device, and the interaction devices. Moreover, a printer and a plotter are also connected to the computer, as shown in that figure.

The display device is usually a CRT. There are two basic types of these devices, namely, refresh and storage CRTs. In the former, the drawing must be refreshed 30 to 60 times per second from the digital representation stored in the graphics data buffer in order to avoid flicker. In the latter, the image is stored as an internal charge distribution, and thus, neither a digital buffer nor a refreshing cycle is necessary.

The display processing unit can be considered as a special-purpose central processing unit. The DPU can be organized to create a drawing either by *random scan* or *raster scan*. In a random scan—also called a *vector, stroke, or calligraphic* system—parts of the drawing can be depicted on the display in any order. In a raster scan—TV-type system—the drawing is divided into horizontal lines. All parts of the drawing appearing in the first line are reproduced in left-to-right order; then, all parts of the drawing in the second line, etc. Common raster graphics systems use anywhere from 256 to 1024 lines. Hard-copy devices also operate with either a random scan or a raster scan. The printer is a simple raster-scan, hard-copy device. The pen plotter, in which a pen can be moved in any direction over a piece of paper—or any other appropriate material—, is a random-scan device.

The resolution of the device is defined as the number of distinguishable pixels per mm (or inch) along the printed line, and along a direction perpendicular to it. A low-resolution printer—similar to almost extinct electrical typewriters—has 10 characters per inch horizontally and 6 characters per inch vertically. Some printers use dot matrix techniques to print discrete dots at resolutions of up to 100 dots per inch. A laser printer has about 300 pixels per inch. This book, including figures, was produced with a laser printer having 300 pixels per inch. Laser technology has now rendered dot-matrix printers obsolete.

Plotters can be classified in the same categories as the displays. Thus, there are random and raster plotters. *Flatbed plotters* draw at random with a pen on a sheet of paper or other medium, spread out on a table and held down by electrostatic charge, either by vacuum, or by stretching. *Drum plotters* are another kind of random plotters. These draw with a pen on a roll of paper stretched tightly across a drum. The most common raster plotter is the *electrostatic plotter*. It works by first depositing a negative charge on those parts of white paper strip that are to be black and then flowing a positively-charged black toner over the paper. The particles adhere to the paper when the charge is deposited.

The advantages of pen plotters over electrostatic plotters are

- they produce sharp, high contrast images, and
- they can use pens of different colors and thicknesses.

The disadvantage of pen plotters is that they are 10 to 20 times slower than electrostatic plotters; however, the latter require scan conversions, which are time consuming.

Among the physical interaction devices frequently found in graphics systems is the *mouse*. It is a hand-held device with rollers on its base. It has several—usually three—push buttons to input commands or other information. When the mouse moves, its relative motion is sensed by potentiometers coupled to one roller. This information concerns the direction and magnitude of the displacements undegone by the mouse. One of the *functions* allowed by the push-buttons, namely, the *pick* function is used to select and move objects around in a drawing, as suggested in Chapters 4 and 5.

9.2.2 Curve-Smoothing Techniques

The rendering of free-form 3-D curves in computer graphics is commonly based on parametric cubic curves. These are chosen because they can provide continuity of position and tangency at the point where curve segments meet and ensure that the ends of the curve segment pass through the specified points. Moreover, their parametric representation allows one to handle curves that cross themselves or that have cusps. In Chapter 8, we discussed the use of cubic, periodic, parametric splines. Here, we present three more ways to define a parametric cubic curve, namely the *Hermite*, the *Bézier*, and the *B-spline* curves. The derivation of the formulas given below is omitted; the interested reader is referred to Foley and Van Dam (1983), Mortenson (1985), or Gardan (1986) for a more detailed discussion. All curves defined below are given in terms of functions of a parameter t, where $0 \leq t \leq 1$.

The Hermite form of a parametric cubic is determined from the end points P_1, P_2 and the end-point tangents as shown in Fig. 9.2.5. Thus, points $P_1(x_1, y_1)$ and $P_2(x_2, y_2)$ and the tangent vectors $R_1 = [x_1', y_1']^T$ and $R_2 = [x_2', y_2']^T$, where $u_i' \equiv du_i/dt$ for $u = x, y$, and $i = 1, 2$, are given.

From the end-point conditions we have:

$$x(0) = x_1, \quad x(1) = x_2, \qquad y(0) = y_1, \quad y(1) = y_2$$

whereas, from the requirement of slope continuity, we have

$$x'(0) = x_1', \quad x'(1) = x_2', \qquad y'(0) = y_1', \quad y'(1) = y_2'$$

Therefore, curves $x(t)$ and $y(t)$ can be represented as

$$x(t) = H_1 x_1 + H_2 x_2 + H_3 x_1' + H_4 x_2' \tag{9.2.1a}$$

$$y(t) = H_1 y_1 + H_2 y_2 + H_3 y_1' + H_4 y_2' \tag{9.2.1b}$$

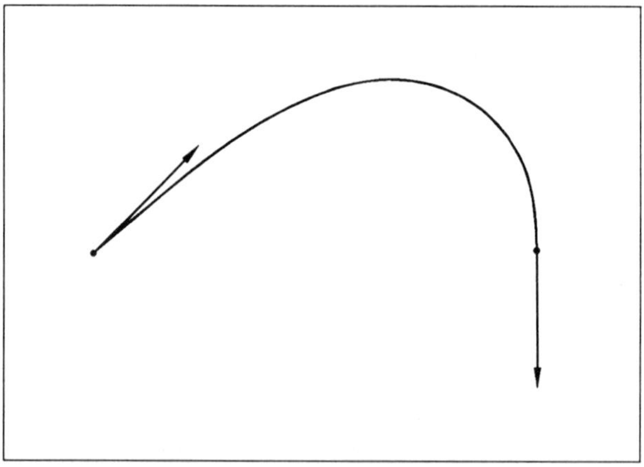

Fig. 9.2.5. The Hermite parametric curve.

where

$$H_1 = 2t^3 - 3t^2 + 1$$

$$H_2 = -2t^3 + 3t^2$$

$$H_3 = t^3 - 2t^2 + t$$

$$H_4 = t^3 - t^2$$

Functions H_i are called *blending functions.*

For the Bézier form, four *control* points, $P_i\,(x_i,\ y_i)$, for $i = 1, \ldots, 4$, are used as shown in Fig. 9.2.6. The tangent vectors of the end points are determined from the line segments $\overline{P_1P_2}$ and $\overline{P_3P_4}$, i.e.,

$$x_1' = 3\,(x_2 - x_1), \quad x_4' = 3\,(x_4 - x_3)$$

with similar expressions for y_i', for $i = 1, 4$. Thus,

$$x(t) = B_1\,x_1 + B_2\,x_2 + B_3\,x_3 + B_4\,x_4 \tag{9.2.2a}$$

$$y(t) = B_1\,y_1 + B_2\,y_2 + B_3\,y_3 + B_4\,y_4 \tag{9.2.2b}$$

where

$$B_1 = (1 - t)^3$$

$$B_2 = 3\,t\,(t - 1)^2$$

$$B_3 = 3\,t^2\,(1 - t)$$

$$B_4 = t^3$$

One important feature of the Bézier form is that the curve thus generated lies within the convex polygon defined by the four given points. This polygon, known as the *convex hull*, bounds the Bézier curve. It should be pointed out that Bézier curves and surfaces can be derived from the Casteljeau construction (Casteljeau, 1985) in a very simple and systematic fashion.

The B-spline cubic representation, in general, does not pass through any of the given points. However, it is continuous and also has continuity of slope and curvature. This is equivalent to saying that the second derivatives are continuous at the end points. Therefore, the B-spline form is smoother than the Hermite and Bézier curves.

Moreover, B-spline curves have the property that each segment is influenced by only j control points, and conversely, each control point influences only j curve segments (Mortenson, 1985).

Concerning smoothing-surface techniques, the simplest mathematical model for modeling a surface is a *patch* (Mortenson, 1985). This is a curve-bounded set of points whose coordinates are given by continuous, two-parameter, single-valued mathematical functions of the form:

$$x = x(u,w), \quad y = y(u,w), \quad z = z(u,w), \qquad 0 \leq u \leq 1, \quad 0 \leq w \leq 1$$

Associated with every patch, there is a set of boundary conditions, namely, the four corner points and the four curves defining its edges. The bicubic patch, the most common one, is defined by the coordinates and the tangent vectors at the corner points. This is often called a *Hermite patch* and is shown in Fig. 9.2.7. For the modeling of complex composite surfaces the most suitable elements are the Bézier and the B-spline surfaces. As in the 2-D case, the counterpart of the convex hull in 3-D is the convex or *characteristic polyhedron*. The mathematical background pertaining to solid modeling is outside the scope of this book. For a comprehensive discussion, the interested reader is referred to the specialized literature—see, for example (Mortenson, 1985).

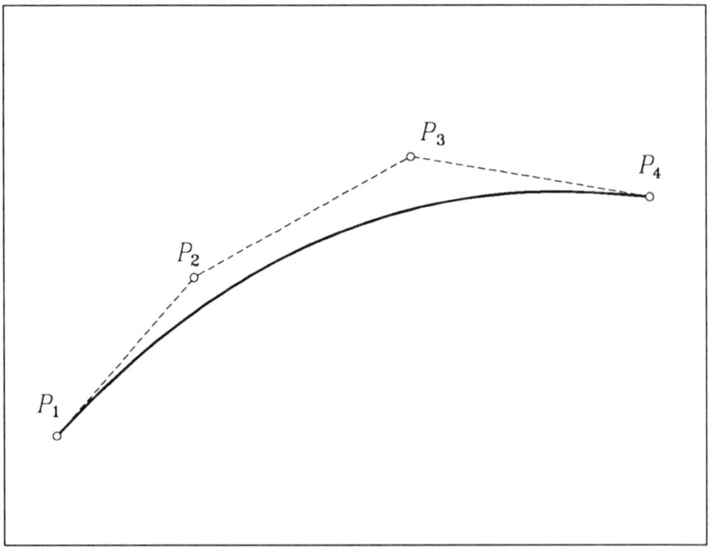

Fig. 9.2.6. The Bézier parametric curve.

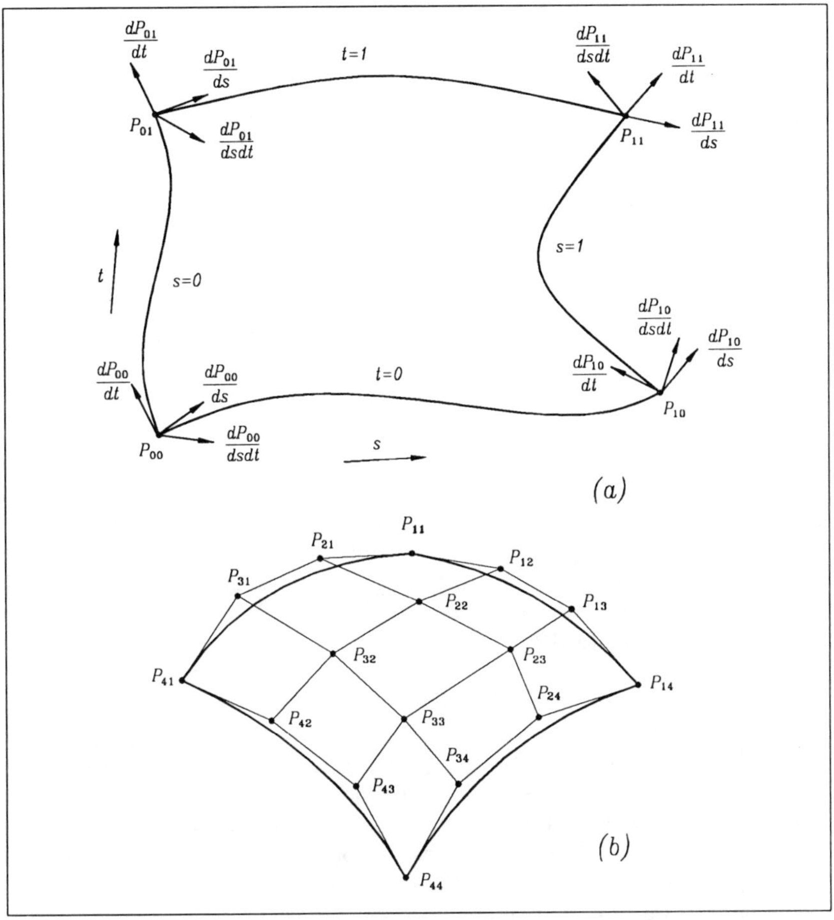

Fig. 9.2.7. The Hermite (a) and the Bézier (b) bicubic patches.

The rendering of realistic surfaces requires *shading* of the different patches. The simplest form of shading is *uniform shading*, which consists of giving each patch a constant light intensity, thereby bringing about intensity discontinuities from patch to patch. This undesirable effect is avoided by introducing a technique known as *intensity interpolation shading* or *Gouraud shading* (Foley and Van Dam, 1983). Most color graphics workstations come with integrated facilities to produce Gouraud shading. Examples of the output produced with the use of this technique appear in the solid models shown in Chapters 4, 5, and 6.

9.2.3 CAD Software Commercially Available

Extensive CAD software is commercially available, with new packages and new features appearing every year. Thus, when planning a CAD facility, the user faces a decision-making task that may require extensive searching and shopping around.

Factors to be taken into account when planning to purchase CAD software are:

- the user's requirements;
- the user's supporting facilities; and, of course,
- the user's budget.

First and foremost, the user must decide whether the planned tasks warrant 3-D software or whether 2-D software will be adequate. The difference in prices is worth considering! Some software is modular and expandable, so the user can buy the basic items and build upon them.

After the user's requirements comes the question of which environment should support the CAD software to be installed. Most popular CAD software was developed for personal computers. Of these, there are two basic types, the difference stemming from the processor. We thus have the Intel 80xxx- series—where xxx stands for 286, 386, or 486—type processors and the Motorola MC68zzz- series—where zzz stands for 000, 020, and 030—processors. The former support the popular DOS (Disk Operating System) used for IBM PCs and compatible machines, the latter are used by the MacIntosh series. Most software suppliers that originally developed packages for one operating system now offer software for the other also. When purchasing software, the user must make sure that it is appropriate for the operating system at hand. Apart from CAD software running on stand-alone personal computers, a variety of packages exist for sophisticated workstations. The differences between graphics workstations lie in their processors. Most common among these are the Motorola 68xxx-processors—where xxx stands for 000, 020, and 030. Motorola processors support the very powerful UNIX operating system. There are various UNIX versions, and software running on one may not run on the others. Some CAD software available for personal computers is also available for UNIX-based workstations. AutoCAD, a trademark of Autodesk Inc., is probably the most widespread software. However, some CAD software meant for UNIX-based workstations is available as a module of a more comprehensive package meant for advanced applications known as MCAE (Mechanical Computer Aided Engineering). Such is the case of GEOMOD, the geometric modeling module of I-DEAS, an integrated package for finite element analysis (FEA) and mechanical analysis of multibody systems. The Intel version of I-DEAS, meant for IBM machines, is available under the name CAEDS. However, the most popular brands of software supported by IBM workstations are CADAM and CATIA. Finally, CISIGRAPH is one of the most comprehensive, and hence, most expensive, pieces of CAD software. In fact, CISIGRAPH is meant for MCAE and contains modules for FEA and computer-aided manufacturing (CAM).

Special-purpose software, although scarce, is also commercially available. Of this kind of software, we can mention the computer package CAM, developed at *Centrum voor het Wetenschappelijk en Technisch Onderzoek der Metaalverwerkende Nijverheid*, in Belgium (De Fraine, 1982). Software for the kinematic analysis, design, and rendering of cam profiles has been developed at the ELITEX Complex Research Institute, in Liberec, Czechoslovakia (Koloc and Václavík, 1988).

The most advanced software allows not only for the geometric modeling of complex mechanical systems comprising elements with complex shapes, but also for the animation of these systems. Because of the tremendous computational requirements involved, the quality of the animation depends, however, on the supporting hardware.

9.3 The Manufacture of Cams

In this section we give a brief review of the conventional methods for cam production. Then, we outline the Computer-Aided Manufacture (CAM) of cams, including commercially available software and hardware.

9.3.1 Conventional Manufacturing Methods

The manufacture of cams has involved many different production methods. For example, machine-shop methods such as milling, grinding, turning, cutting, sawing, etc., are those most commonly considered when the manufacture of cams is discussed. However, if large batches of cams have to be produced, industrial type methods such as casting, molding, punching, sintering, stamping, extruding, etc., are more likely to be used (Grant and Sony, 1979). The production methods of master cams and copied cams are briefly outlined below. The interested reader is referred to the aforementioned reference for a more detailed discussion.

Original or Master Cams

These cams are produced from the synthesis equations discussed in Chapters 4 and 5 or from graphs. There are basically three methods for the production of the master cams, namely,

- layout methods,
- increment methods, and
- contact-rise and circular-arc cams.

Layout Methods

These are quick and simple, but lack accuracy. The first step in this procedure consists of laying out or scribing the desired profile on the cam blank. The second step, i.e., the actual manufacture of the cam, can be accomplished with either a contour cutting bandsaw—and for this reason these methods are called *bandsaw methods*—or grinding machines, and mills with grinding heads. All these methods are usually limited to larger cams of the radial type where accuracy is not important. Moreover, accuracy is highly dependent on the machine operator. Tolerances of 0.075 mm to 0.1 mm can be obtained with the use of a good manufacturing device and the work of an experienced operator, but accuracies of 0.25 mm are more common.

Increment Methods

These are much slower but more accurate for the production of original cams—they are meant for later production of copied cams. The procedure consists of plotting many points, using small increments of the angle of rotation of the cam, and computing the corresponding polar coordinates that describe the cam profile on the cam blank. Then, a mill head, which in the majority of cases must be of the same radius as the cam roller-follower, is driven through all the points marked on the blank. After one cuts the cam in this way, the cam perimeter requires hand finishing to eliminate the cusps and to leave a smooth contour. Tolerances of the order of ± 0.0127 mm can be obtained with an experienced operator and many increments. The most common pieces of machine shop equipment that are used in conjunction with the increment techniques are mills with a rotary table and dividing head.

Methods Applicable to Constant-Rise and Circular-Arc Cams

The profiles of a large class of disk cams consist of a combination of simple planar primitives, namely, lines and circular arcs, which can be machined efficiently and accurately using basic operations like shaping and turning. The manufacturing methods applicable to these cams are therefore more accurate and quicker than the previously mentioned methods. However, these cams are dynamically unsatisfactory for the reasons discussed in Chapters 2, 4, and 5.

9.3.2 Production Methods for Copied Cams

The most common machine shop methods for producing copied cams can be classified in two categories. The first category involves the use of commercially available attachments for common machine tools such as horizontal and vertical mills, lathes, and others. These attachments allow smaller job shops to produce cams with less machine operator dependence. An example of this type of attachments is the *cam slide* that allows the cutting of duplicate or copied cams on a production basis. In the second category, cams are produced with special machine tools such as mills or lathes, with which the cam blank is cut, ground, or turned to the desired profile. A *pantograph profiling machine* is an example of a mechanical system used to copy cams from an original template or master cam.

9.3.3 NC Manufacturing Methods

CAD methods of producing a database describing the cam profile require the use of a numerically-controlled (NC) machine tool to produce the cam. In order to describe the sequence of operations to be performed by an NC machine tool for the manufacturing of a desired workpiece, one has to resort to a language for the programming of the various operations. This set of instructions is referred to as the *part-program*.

In general, the steps described below are required to generate a control tape or a program for driving an NC machine tool (Besant, 1983):

- The *part-programmer* is given an engineering drawing of the required part. From this, a *part-program* manuscript is produced, which involves the coding of the geometry of the item followed by the coding of statements to describe the tool motions required to produce the machining.
- The manuscript is input into an NC processor, and, after debugging, a control file is produced.

One of the most common languages for describing the sequence of operations to a machine tool is APT—for Automatically Programmed Tools (Besant, 1983). It is both a language and a computer program.

APT is capable of allowing a user to produce complex three-dimensional geometrical contours of tool motion from a complex three-dimensional design. The APT system consists of three parts, namely, the APT part-program, the APT program processor, and the APT program post-processor. The part-program is a set of instructions written in English-like statements which describe the tool and its motions. This part-program is input into the main program, which in turn converts the simple instructions of the part-program into a set of numerical commands by performing the necessary computations. These commands are then processed by a part-processor program. The APT processor is a very large program, normally residing on a large storage unit which is run mostly on a mainframe. Once the part-program is input, the APT processor produces an output which is called the *cutter location data*—CLDATA. These data are then normally passed on to an APT post-processor that produces a numerical-control file for a specific machine tool.

One of the advantages of APT is that it has become a worldwide standard for NC machines. On the other hand, one of the disadvantages in the use of the APT system is the large overhead in terms of computing capacity and power for relatively simple components. In order to avoid this burden, APT can be used via a time-sharing system, but the tradeoff is the slow system response when used in this mode. Another disadvantage of the APT system is its dependence on the availability of a post-processor for the machine tool at hand; in the absence of the post-processor, the control files for that machine cannot be produced.

Some variations of APT are listed below:

- EXAPT, developed in Germany;
- IFAPT, developed in France;
- NELAPT, developed in the United Kingdom; and
- ICAM, developed in Canada.

In EXAPT—the acronym for EXtended Automatically Programmed Tools—, for example, a program complying with the ISO-Code and the DIN 66024 and DIN 66025 norms is written in the form of statements. A statement is composed of words, each statement representing elementary instructions for the NC-machine. Each word, in turn, represents an address, which starts with a letter followed by some numerical characters. A common statement involves: the statement number; the operation conditions; the point coordinates; the feed speed; the cutting speed; and some supplementary functions. The program is stored in files with eight tracks, seven of which contain the alphanumeric code.

EXAPT has three main versions (Weck, 1978), depending upon the operation at hand: EXAPT1, for boring; EXAPT2, for turning; and EXAPT3, for milling. The CLDATA2 file is machine-dependent. It is produced in an intermediate step, which is executed by a machine-specific post-processor. For example, it verifies if the values obtained for the cutting speed and feed speed can be performed in the selected machine tool. Although we do not provide similar details for IFAPT, this language should not be too different from EXAPT.

As to NELAPT, it can run on large workstation as well as on mainframes. In fact, the minimum hardware configuration on which the NELAPT system can be implemented consists of:

- 28 Kbytes of 24-bit words of primary memory;
- four magnetic tapes and disk storage;
- one input device such as a card reader or a computer terminal;
- one line printer; and
- one output device such as a diskette.

ICAM Technologies (ICAM, 1985) released a software called CAM-POST, which consists of two parts. The first part, CAM-QUEST, builds a database pertaining to an NC machine tool. The second part, CAM-GENER, uses this information to create a punched tape for a specific machine, or any other compatible CLDATA file.

Furthermore, the CAM package mentioned before (De Fraine, 1982) and the ELITEX software (Koloc and Václavík, 1988) are supplied with cam-oriented NC capabilities.

9.3.4 Computation of Tool Trajectories

We focus on the machining of cam plates and assume that a circular mill is used as a cutting tool on an NC machine tool. Thus, the trajectory of the mill center, henceforth referred to as the cutter trajectory, should be given in a database, which consists of a set of N Cartesian coordinate pairs $(x_k,\ y_k)$ on the trajectory. If the cam mechanism at hand has a roller-follower, then the cutter trajectory is computed using the procedures described in Chapter 5 for a knife-edge follower. As discussed in that chapter, the roller radius must be selected so as to avoid undercutting. If the roller radius is, additionally, selected as that for the cutting mill, then the cam profile obtained for the knife-edge follower is identical to the cutter trajectory. Fig. 9.3.1 shows a cam plate with a roller follower and the cutter trajectory.

For flat-face followers, we select a cutter so that undercutting is avoided. We do this by specifying the cutter radius not smaller than that of the minimum radius of curvature of the cam profile, r_m. We now assume that the cutter has a radius of $a < r_m$. The cutter trajectory is then computed from the geometry of Fig. 9.3.1.

From Fig. 9.3.1, the position vector $\mathbf{c}$ of the cutter center C is found from that of P, denoted by $\mathbf{p}$, and the unit normal $\mathbf{e}_n$, i.e.,

$$\mathbf{c} = \mathbf{p} - a\,\mathbf{e}_n \tag{9.3.1}$$

Vector $\mathbf{p}$ is available in the form $\mathbf{p} = x(\psi)\,\mathbf{i} + y(\psi)\,\mathbf{j}$, where $\mathbf{i}$ and $\mathbf{j}$ are unit vectors parallel to the coordinate axes. Moreover, vector $\mathbf{e}_n$ is computed as

$$\mathbf{e}_n = \mathbf{E}\,\mathbf{e}_t$$

where the 2×2 matrix $\mathbf{E}$ was introduced in Chapter 3 and defined as

$$\mathbf{E} = \begin{bmatrix} 0 & -1 \\ 1 & 0 \end{bmatrix}$$

Furthermore, $\mathbf{e}_t$ is the unit tangent vector of the cam profile and is determined as

$$\mathbf{e}_t = \frac{\mathbf{p}'(\psi)}{\|\mathbf{p}'(\psi)\|} \tag{9.3.2}$$

For translating flat-face followers, $\mathbf{p}'(\psi)$ was already derived in eq. (4.2.28d), namely,

$$\mathbf{p}'(\psi) = \mathbf{Q}^T(\psi)\,[\mathbf{r}'(\psi) - \mathbf{E}\,\mathbf{r}(\psi)] \tag{9.3.3}$$

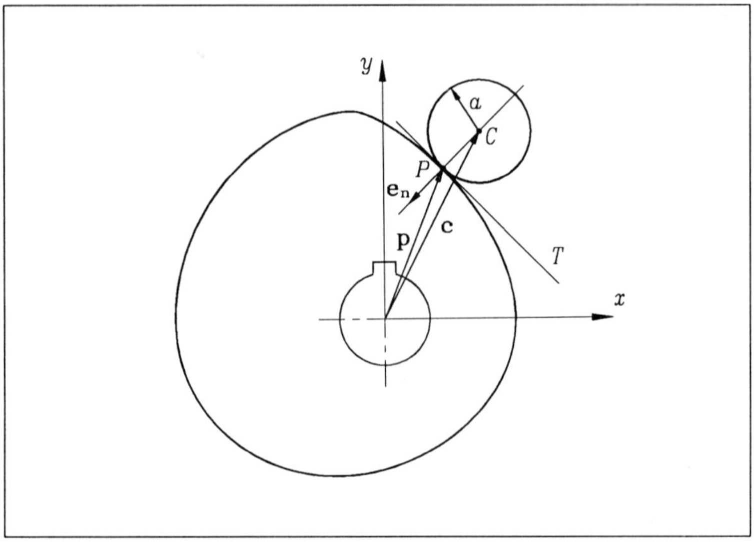

Fig. 9.3.1. Cam plate with a roller follower and the cutter trajectory.

where $\mathbf{r}(\psi)$ is given in terms of the follower displacement program $s(\psi)$ and its derivative $s'(\psi)$ in eq. (4.2.9c), while $\mathbf{r}'(\psi)$ is given in eq. (4.2.30) in terms of the first two derivatives of the displacement. Thus,

$$\mathbf{r}'(\psi) - \mathbf{E}\,\mathbf{r}(\psi) = \begin{bmatrix} s''(\psi) \\ s'(\psi) \end{bmatrix} - \begin{bmatrix} 0 & -1 \\ 1 & 0 \end{bmatrix} \begin{bmatrix} s'(\psi) \\ s(\psi) \end{bmatrix} = \begin{bmatrix} s''(\psi) + s(\psi) \\ 0 \end{bmatrix} \tag{9.3.4}$$

Substitution of eq. (9.3.4) into eq. (9.3.3) yields

$$\mathbf{p}'(\psi) = \begin{bmatrix} \cos\psi & \sin\psi \\ -\sin\psi & \cos\psi \end{bmatrix} \begin{bmatrix} s''(\psi) + s(\psi) \\ 0 \end{bmatrix} = [\,s''(\psi) + s(\psi)\,] \begin{bmatrix} \cos\psi \\ -\sin\psi \end{bmatrix} \tag{9.3.5}$$

and hence,

$$\|\,\mathbf{p}'(\psi)\,\| = |\,s''(\psi) + s(\psi)\,| \tag{9.3.6}$$

Moreover, if the design was verified for cusp avoidance, the cam profile thus obtained will be convex, i.e., the curvature will be positive everywhere.

Upon substituting eqs. (9.3.5) and (9.3.6) into eq. (9.3.2), one obtains $\mathbf{e}_t$ as

$$\mathbf{e}_t = \begin{bmatrix} \cos\psi \\ -\sin\psi \end{bmatrix} \tag{9.3.7}$$

Therefore,

$$\mathbf{e}_n = \mathbf{E}\,\mathbf{e}_t = \begin{bmatrix} \sin\psi \\ \cos\psi \end{bmatrix}$$

Now, vector $\mathbf{c}$ can be computed from eq. (9.3.1), (4.2.10a & b), and the foregoing expression for $\mathbf{e}_n$ as

$$\mathbf{c} = \begin{bmatrix} [\,s(\psi) - a\,]\sin\psi + s'(\psi)\cos\psi \\ [\,s(\psi) - a\,]\cos\psi - s'(\psi)\sin\psi \end{bmatrix} \tag{9.3.8}$$

which is the relation sought.

In the case of *oscillating* flat-face followers, $\mathbf{p}'(\psi)$ is obtained from eq. (4.3.29a) as

$$\mathbf{p}'(\psi) = u \begin{bmatrix} \cos\mu \\ \sin\mu \end{bmatrix} \tag{9.3.9}$$

where u is defined as

$$u \equiv \bar{\xi}'(\psi) + \mu'(\psi)\,\bar{\eta}$$

while $\mu'(\psi)$ and $\bar{\xi}'(\psi)$ were already computed in Subsection 4.3 and are repeated here for quick reference, namely,

$$\mu'(\psi) = 1 + \phi'(\psi)$$

$$\bar{\xi}'(\psi) = \frac{\phi'^2(1+\phi')\sin\phi + \phi''\cos\phi}{[\,1+\phi'(\psi)\,]^2}$$

Now,

$$\|\,\mathbf{p}'(\psi)\,\| = |\,u\,|$$

As for translating flat-face followers, the curvature will be greater than zero. Thus, $\mathbf{e}_t$ is given as

$$\mathbf{e}_t = \frac{\mathbf{p}'(\psi)}{\|\,\mathbf{p}'(\psi)\,\|} = \begin{bmatrix} \cos(\psi+\phi) \\ -\sin(\psi+\phi) \end{bmatrix} \tag{9.3.10}$$

where the definition of μ given in Subsection 4.3, namely, $\mu \equiv \psi + \phi$, has been recalled, while $\mathbf{e}_n$ is then computed as

$$\mathbf{e}_n = \mathbf{E}\,\mathbf{e}_t = \begin{bmatrix} \sin(\psi+\phi) \\ \\ \cos(\psi+\phi) \end{bmatrix}$$

Finally, for a given normalized cutter radius $\bar{a}\,(\equiv a/\ell)$, $\mathbf{c}$ is given as

$$\mathbf{c} = \begin{bmatrix} \frac{\phi'(\psi)\cos\phi}{1+\phi'(\psi)}\cos(\psi+\phi) + (\sin\phi + \bar{e} - \bar{a})\sin(\psi+\phi) \\ -\frac{\phi'(\psi)\sin\phi}{1+\phi'(\psi)}\sin(\psi+\phi) + (\sin\phi + \bar{e} - \bar{a})\cos(\psi+\phi) \end{bmatrix} \tag{9.3.11}$$

thereby completing the computation of the trajectory of the cutting center for this type of disk cam.

9.3.5 Computer Integration of the Design and Manufacture of Cams

To conclude this work we describe the information flow for the computer integration of the design and manufacture of cams, shown in Fig. 9.3.2. In the first block, the designer inputs either the full displacement program or the percentages of the upper and lower dwells, as well as those of the rise and return phases. In the block labeled OPTCAM, the optimum parameters of the cam mechamism, the Cartesian coordinates of the cam profile and its inertia properties are computed for a given follower-displacement program and prescribed functional constraints. As a result, if 3-D software modeling is available, a solid model of the cam profile is displayed as shown in the same figure; otherwise, only a 2-D model is displayed. This model can be generated by some of the software packages mentioned in Section 9.2.3. It is pointed out that, at this stage, designers have the option of going back to the previous stage and modifying their designs, if necessary. Next, the NC code for the cutting of the cam profile is generated in the GCAM block, e.g., in the form of a G program. This code is input into an NC milling machine, shown in block 4, where the actual profile is manufactured.

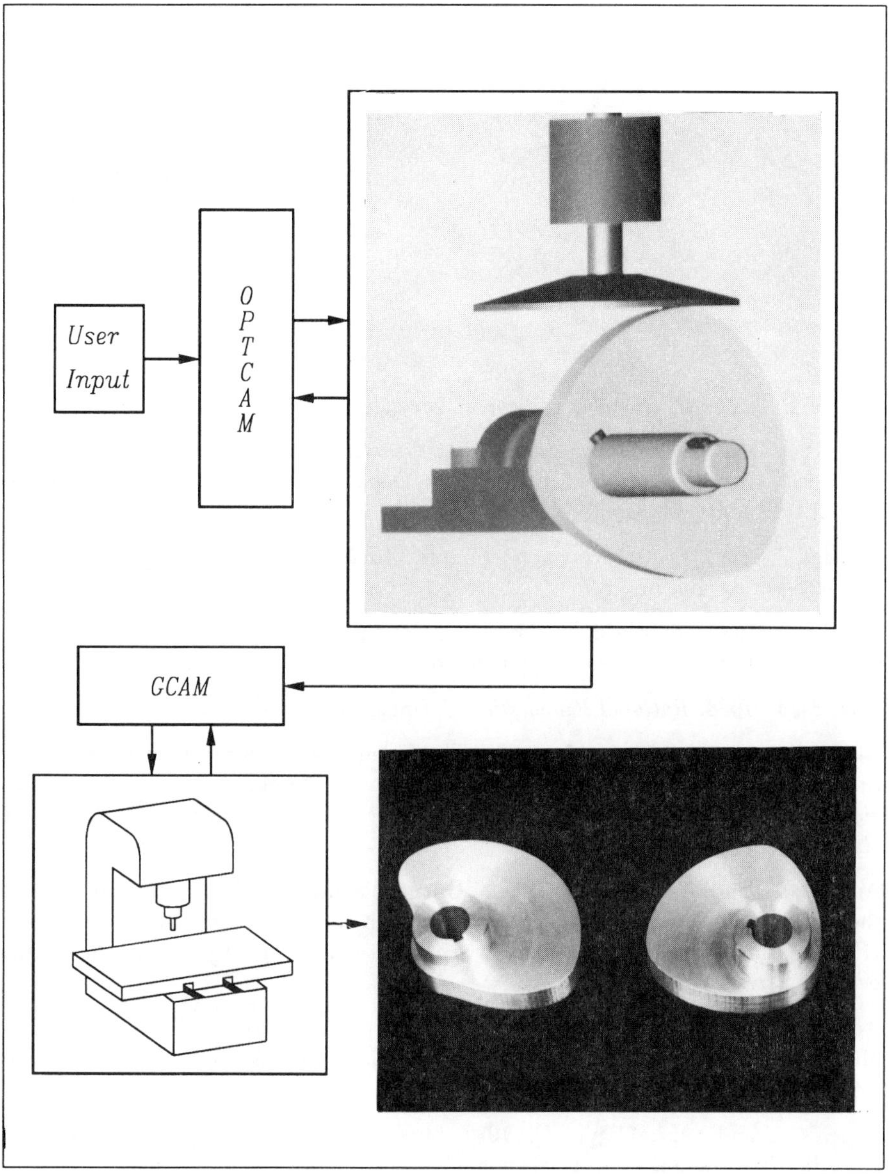

Fig. 9.3.2. Information flow.

REFERENCES

Adams, R. A., 1987, *Calculus of Several Variables*, Addison-Wesley, Reading.

Al-Daccak, M., 1989, *Contributions to the Geometric Modelling of Mechanical Elements with Complex Shapes*, M. Eng. Thesis, Department of Mechanical Engineering, McGill University, Montreal.

Angeles, J., 1982, *Spatial Kinematic Chains. Analysis, Synthesis and Optimization*, Springer-Verlag, Berlin.

Angeles, J., 1983, "Synthesis of Plane Curves with Prescribed Local Geometric Properties using Periodic Splines", *Computer Aided Design*, Vol. 15, No. 3, pp. 147–155.

Angeles, J., 1988, *Rational Kinematics*, Springer-Verlag, New York.

Angeles, J, and Al-Daccak, M., 1989, "The Evaluation of Moments of Regions with Piecewise-Linearly Approximated Boundaries via Line Integration", *Advances in Design and Automation-1989. Volume One: Computer-Aided and Computational Design*, DE-Vol. 19–1, pp. 163–168.

Angeles, J. and Arteaga, O., 1977, "Optimal synthesis of cam mechanisms via the methods of Newton-Raphson and Runge-Kutta", *Proceedings of the Second IFToMM International Symposium on Linkages and Computer-Aided Design Methods*, Bucharest, Vol III-1, pp. 1–12.

Angeles, J., Chaoming, Z., Duarte, R. and Ning, H., 1990, "The Evaluation of Moments of 3-D Solids of Revolution Using Spline Approximations of the Boundary", *ASME Journal of Mechanical Design* Vol. 112, pp. 108–110.

Angeles, J. and López-Cajún, C., 1983, "Diseño Automatizado de Mecanismos de Leva de Disco con Seguidor Traslacional de Cara Plana", *Memoria del IX Congreso de la ANIAC*, León, Guanajuato, México, pp. 91–93.

Angeles, J. and López-Cajún, C., 1984a, "Optimal Synthesis of Translating Roller-Follower Cam Mechanisms with Prescribed Functional Constraints", *Proceedings of the First International Symposium on Design and Synthesis*, Tokyo, Japan, pp. 782–787.

Angeles, A. and López-Cajún, C., 1984b, "Optimal Synthesis of Oscillating Roller-Follower Cam Mechanisms with Prescribed Functional Constraints", *Proceedings of the ASME International Conference on Computers in Engineering*, Las Vegas, Nevada, Vol. 2, pp. 273–277.

Angeles, J. and López-Cajún, C., 1988, "Optimal Synthesis of Cam Mechanisms with Oscillating Flat-Face Followers". *Mechanism and Machine Theory*, Vol. 23, No. 1, pp. 1–6.

Angeles, J., Chaoming, Z., Duarte, R., and Ning, H., 1988, "The Evaluation of Moments of Bounded Regions Using Spline Approximations of the Boundary", *Proceedings of the 1988 ASME Design Automation Conference*, Kissimmee, Florida, DE-Vol. 14, pp. 49–54.

Angeles, J., López-Cajún, C., and Saha, S. K., 1989a, "Curvature Constraints in the Design of Cam Mechanisms. Part I: Translating Flat-Face Followers", *Advances in Design Automation-1989. Volume Three: Mechanical Systems, Design, and Simulation*, DE-Vol. 19-3, Montreal, Canada, pp. 259–264.

Angeles, J., López-Cajún, C., and Saha, S. K., 1989b, "Curvature Constraints in the Design of Cam Mechanisms. Part II: Oscillating Flat-Face Followers", *Advances in Design Automation-1989. Volume Three: Mechanical Systems, Design, and Simulation*, DE-Vol. 19-3, Montreal, Canada, pp. 265–269.

Archbold, J. W., 1970, *Algebra*, Fourth Ed., Pitman Publishing, London.

Backhouse, C. J. and Rees Jones, J., 1989, "Cam and Follower Contact Properties in Globoidal Indexing Mechanisms", *Vortragssammelband II IFToMM-Symposium Kurvengetriebe*, Berlin, pp. 84–95.

Beggs, J. S., 1959, *Ein Beitrag zur Analyse Räumlicher Mechanismen*, Doctoral Thesis, Technische Hochschule Hannover, Hannover.

Berzak, N., 1982, "Optimization of Cam-Follower Systems with Kinematic and Dynamic Constraints", *ASME Journal of Mechanical Design*, Vol. 104, pp. 29–33.

Besant, C. B., 1980, *Computer-Aided Design and Manufacture*, Ellis Horwood Ltd., Chichester.

Brand, L., 1965, *Advanced Calculus*, John Wiley & Sons, Inc., New York.

Bryson, A. and Ho, Y., 1975, *Applied Optimal Control: Optimization, Estimation, and Control*, Hemisphere Pub. Co., Washington.

Buchsbaum, F. and Freudenstein, F., 1983, "On a Class of Cam-Type Angular Motion Compensators", *ASME Journal of Engineering for Industry*, Vol. 85, No. 2, pp. 487–496.

Burr, A., 1981, *Mechanical Analysis and Design*, Elsevier Science Publishers, B. V., Amsterdam.

Cannon, R. H., 1967, *Dynamics of Physical Systems*, McGraw-Hill Book Co., New York.

de Casteljeau, P. 1985, *Mathematics and CAD*, Vol. 2, Kogan Page, London.

Chakraborty, J. and Dhande, S. G., 1977, *Kinematics and Geometry of Planar and Spatial Cam Mechanisms*, John Wiley, New York.

Chen, F. Y., 1977, "A Survey of the State of the Art of Cam Systems Dynamics", *Mechanism and Machine Theory*, Vol. 12, pp. 201–224.

Chen, F. Y., 1982, *Mechanics and Design of Cam Mechanisms*, Pergamon Press, New York.

Chicurel, R., 1963, "Cam Size Minimization by Offsetting", *ASME Paper No. 62-WA-22*.

Cowie, A., 1961, *Kinematics and Design of Mechanisms*, International Textbook Co. Scranton Pennsylvania.

Dahlquist, G. and Björck, Å. 1974. *Numerical Methods*, Prentice-Hall, Inc., Englewood Cliffs.

De Fraine, J., 1982, "The Computer-Aided Design and Computer-Aided Manufacturing of Cams", Technical Report, Centrum voor Wetenschappelijk en Technisch Onderzoek der Metaalverwerkende Nijverheid, MC 75, November.

Dittrich, G., 1966, "Konstruktion und Herstellung Sphärischer Kurvengetriebe", *Konstruktion*, Vol. 18, No. 10, pp. 421–426.

Dittrich, G., 1970, "Vergleich von Ebenen, Sphärischen und Räumlichen Getrieben", *VDI-Berichte*, No. 140, pp. 25–34.

Dudiţă, Fl., Gogu, Gr., Diaconescu, D., and Stareţu, I., 1989, *Curs de Mecanisme. Fascicula 2. Angrenaje. Mecanisme cu Cama. Cinematica*, University of Braşov, Braşov.

Foley, J. D. and Van Dam, A., 1983, *Fundamentals of Interactive Computer Graphics*, Addison-Wesley Publishing Co., Reading.

Forsythe, G. Malcolm, M. A., and Moler, C. B., 1977, *Computer Methods for Mathematical Computations*, Prentice-Hall, Inc., Englewood Cliffs.

Gardan, Y., 1986, *Numerical Methods for CAD*, The MIT Press, Cambridge, Massachusets.

Gelfand, I. M. and Fomin, S. V., 1963, *Calculus of Variations*, Prentice-Hall, Inc., Englewood Cliffs.

Goetz, D. and Morley, S. G., 1954, *The Book of the People: Popol Vuh. The National Book of the Ancient Quiché Maya* (English translation from the Spanish version by A. Recinos), The Plantin Press, Los Angeles.

Golub, G. H. and Van Loan, C. F. 1983. *Matrix Computations*, The Johns Hopkins University Press, Baltimore.

Grant, B. and Soni, A. H., 1979, "A Survey of Cam Manufacture Methods", *ASME Journal of Mechanical Design*, Vol. 101, pp. 455–464.

Guoxun, P., Zhengyang, X., and Huimin, T., 1988, "Unified Optimal Design of External and Internal Parallel Indexing Cam Mechanisms", *Mechanism and Machine Theory*, Vol. 23, No. 4, pp. 313–318.

Hartenberg, R.S. and Denavit, J., 1964. *Kinematic synthesis of linkages*, McGraw-Hill, New York.

Hitoshi, T., 1976, "Kinematic Design of Cam Follower Systems", Ph.D. Thesis, Columbia University, New York.

Hunt, K. H., 1978, *Kinematic Geometry of Mechanisms*, Clarendon Press, Oxford.

Hurley, J. F., 1981, *Multivariable Calculus*, Saunders College Publishing, Philadelphia.

ICAM, 1985, *User's Guide*, ICAM Technologies Corporation, Pointe Claire, Quebec, Canada.

Jeffrey, A., 1989, *Mathematics for Engineers and Scientists*, Fourth Ed., Van Nostrand-Reinhold, London.

Johnson, R. C., 1980, *Optimum Design of Mechanical Elements*, John Wiley & Sons, Inc., New York.

Koloc, Z. and Václavík, M., 1988, *Vačkové Mechanismy*, SNTL/ALFA, Prague.

Lee, Y. T. and Requicha, A. G., 1982a, "Algorithms for Computing the Volume and Other Integral Properties of Solids. I. Known Methods and Open Issues", *Communications of the ACM*, Vol. 25, No. 9, pp. 635–641.

Lee, Y. T. and Requicha, A. G., 1982b, "Algorithms for Computing the Volume and Other Integral Properties of Solids. II. A Family of Algorithms Based on Representation Conversion and Cellular Approximation", *Communications of the ACM*, Vol. 25, No. 9, pp. 642–650.

Lelong-Ferrand, J. and Arnaudiès, J-M., 1978,*Cours de Mathématiques, Tome 1, Algèbre*, 3e édition, Dunod, Paris.

Lien, S. and Kajiya, J. T., 1984, "A Symbolic Method for Calculating the Integral Properties of Arbitrary Nonconvex Polyhedra", *IEEE Computer Graphics and Applications*, Vol. 4, pp. 35–41.

Makino, H., 1976, *Mechanisms of Automation Machines*, Japanese Industrial Press, Tokyo.

MATLAB, 1990, *MATLAB for Sun Workstations. User's Guide*, The MathWorks, Inc., South Natick (MA).

Messner, A. M. and Taylor, G. Q., 1980, "Algorithm 550. Solid Polyhedron Measures [Z]", *ACM Trans. Math. Software*, Vol. 6, No. 1, pp. 121–130.

Miles, R. G. and Tough, J. G., 1983, "A Method for the Computation of Inertial Properties for General Areas", *Computer-Aided Design*, Vol. 15, pp. 196–200.

Mischke, C., 1970, "Optimal Offset on Translating Follower Plate Cams", *ASME Journal of Engineering for Industry*, Vol. 92, No. 3, pp. 172–176.

Mortenson, M. E., 1985, *Geometric Modeling*, John Wiley & Sons, New York.

Müller, J., 1987, "The history of Cams and Cam Mechanisms", *Proc. 7th World Congress on Theory of Machines and Mechanisms*, Seville, pp. 1649–1652.

Müller, J. and Kassamanian, A. A., 1979, "Zur Systematik des Kurvengetriebes", *Maschinenbautechnik*, Vol. 28, No. 9, pp. 427–428.

Müller, J. and Mauersberger, K., 1989, "Kurbengetriebe bei Leonardo da Vinci–Ein Technik-Historischer Exkurs in die Renaissance", *Vortragssammelband II IFToMM-Symposium Kurvengetriebe*, Berlin, pp. 3–10.

Nó Sánchez, M. and García de Jalón, J., 1980, "Application of B–Spline Functions to the Motion Specification of Cams", *ASME Paper No. 80-DET-28.*

Ota, Y., Arai, H., and Tokumasu, S., 1985, "An Automated Finite Polygon Division Method for 3-D Objects", *IEEE Computer Graphics and Applications*, Vol. 5, pp. 60–70.

Prenzel, W. D., 1989, "Einsatz Raumlicher Kurvengetriebe in Photographischen Objectiven", *Vortragssammelband II IFToMM-Symposium Kurvengetriebe*, Berlin, DDR, pp. 96–101.

Rees Jones, J., (Editor), 1978, *Cams and Cam Mechanisms*, Mechanical Engineering Publications Ltd., London-Birmingham (Alabama).

Rothbart, H. A., 1956, *Cams. Design, Dynamics and Accuracy*, John Wiley, New York.

Salmon, R. and Slater, M., 1987, *Computer Graphics: Systems and Concepts*, Addison-Wesley Publishing Co., Workingham.

Sermon, C. F. and Liniecki, A., 1972, "Search for Optimum Solution of a Single Disk Cam Mechanism with an Oscillating Roller Follower", *ASME Paper No. 72-MECH-61.*

Späth, H., 1978, *Spline-Algorithmen zur Konstruktion glatter Kurven und Flächen*, 2nd. Edition, R. Oldenburg Publishers, Munich–Viena.

Suh, C. H. and Radcliffe, C. W., 1978, *Kinematics and Mechanisms Design*, John Wiley & Sons Inc., New York.

Terauchi, Y. and El-Shakery, S. A., 1983, "A Computer-Aided Method for Optimum Design of Plate Cam Size Avoiding Undercutting and Separation Phenomena-I", *Mechanism and Machine Theory*, Vol. 18, pp. 157–163.

Tesar, D. and Matthew, G. K., 1976, *The Dynamics, Synthesis, Analysis and Design of Modeled Cam Systems*, Lexington Books, Lexington (Massachusetts).

Thompoulos, N. T. and Knowles, T. W., 1975, " Use of Linear Programming for Cam Design", *Int. Journal of Mach. Tool Des. & Res.* Vol. 15, pp. 257–265.

Timmer, H. G., and Stern, J. M., 1980, "Computation of Global Geometric Properties of Solid Objects", *Computer–Aided Design*, Vol. 12, No. 6, pp. 301–304.

Volmer, J., 1976, *Getriebetechnik. Kurvengetriebe*, VEB Verlag Technik, Berlin.

Weck, M., 1978, *Werkzeugmaschinen, Band 3, Automatisierung und Steuerungstechnik*, VDI-Verlag GmbH, Düsseldorf.

Wolfram, W., 1988, *Mathematica. A System for Doing Mathematics by Computer*, Addison-Wesley Publishing Co., Redwood City.

Woodwark, J. R. and Quinlan, K. M., 1982, "Reducing the Effect of Complexity on Volume Model Evaluation", *Computer-Aided Design*, Vol. 14, pp. 89–95.

Wunderlich, W., 1971, "Contributions to the Geometry of Cam Mechanisms with Oscillating Followers", *Journal of Mechanisms*, Vol. 6, pp. 1–20.

APPENDIX A POLYNOMIAL COEFFICIENTS

This appendix contains the polynomial coefficients appearing in the formulas derived in Chapter 8 for the computation of the geometric properties of cams. For ease of presentation, the subscript i of the spline coefficients, $A_{xi}, \ldots, D_{yi}$, is omitted throughout.

We display the polynomial coefficients of eqs. (8.4.12h–k) in the tables below:

Table A.1

a_{ij}	Expression
a_{i1}	$\frac{1}{2}(C_y D_x - C_x D_y)$
a_{i2}	$\frac{1}{2}(B_y D_x - B_x D_y)$
a_{i3}	$\frac{1}{6}(3A_y D_x - 3A_x D_y + B_y C_x - B_x C_y)$
a_{i4}	$\frac{1}{4}(A_y C_x - A_x C_y)$
a_{i5}	$\frac{1}{10}(A_y B_x - A_x B_y)$

Table A.2

q_{xij}	Expression
q_{xi1}	C_yT_7
q_{xi2}	$C_yT_6 + B_yT_7$
q_{xi3}	$\frac{1}{2}(C_yT_4 + B_yT_5 + 3A_yT_6)$
q_{xi4}	$\frac{1}{3}(C_yT_5 + 4B_yT_6 + 3A_yT_7)$
q_{xi5}	$\frac{1}{5}(C_yT_3 + 4B_yT_4 + 3A_yT_5)$
q_{xi6}	$\frac{1}{3}(C_yT_2 + B_yT_3 + 3A_yT_4)$
q_{xi7}	$\frac{1}{7}(C_yT_1 + 4B_yT_2 + 3A_yT_3)$
q_{xi8}	$\frac{1}{4}(B_yT_1 + 3A_yT_2)$
q_{xi9}	$\frac{1}{3}A_yT_1$

Table A.3

T_i	Expression
T_1	$A_x^2 + A_y^2$
T_2	$A_xB_x + A_yB_y$
T_3	$B_x^2 + B_y^2 + 2(A_xC_x + A_yC_y)$
T_4	$A_xD_x + A_yD_y + B_xC_x + B_yC_y$
T_5	$C_x^2 + C_y^2 + 2(B_xD_x + B_yD_y)$
T_6	$C_xD_x + C_yD_y$
T_7	$D_x^2 + D_y^2$

Table A.4

$H_i^{(1)}$	Expression
$H_1^{(1)}$	$C_x D_x T_7$
$H_2^{(1)}$	$2C_x D_x T_6 + Q_x T_4$
$H_3^{(1)}$	$C_x D_x T_5 + 2Q_x T_6 + 3B_x C_x T_7$
$H_4^{(1)}$	$2C_x D_x T_4 + Q_x T_5 + 6(B_x C_x + A_x D_x)T_6 + 2P_x T_7$
$H_5^{(1)}$	$C_x D_x T_3 + 2Q_x T_4 + 3B_x C_x T_5 + 4P_x T_6 + 5A_x B_x T_7$
$H_6^{(1)}$	$2C_x D_x T_2 + Q_x T_3 + 6B_x C_x T_4 + 2P_x T_5 + 10A_x B_x T_6 + 3A_x^2 T_7$
$H_7^{(1)}$	$C_x D_x T_1 + 2Q_x T_2 + 3B_x C_x T_3 + 4P_x T_4 + 5A_x B_x T_5 + 6A_x^2 T_6$
$H_8^{(1)}$	$Q_x T_1 + 6(B_x C_x + A_x D_x)T_2 + 2P_x T_3 + 10A_x B_x T_4 + 3A_x^2 T_5$
$H_9^{(1)}$	$3B_x C_x T_1 + 4P_x T_2 + 5A_x B_x T_3 + 6A_x^2 T_4$
$H_{10}^{(1)}$	$2P_x T_1 + 10A_x B_x T_2 + 3A_x^2 T_3$
$H_{11}^{(1)}$	$5A_x B_x T_1 + 6A_x^2 T_2$
$H_{12}^{(1)}$	$A_x^2 T_1$

Table A.5

$H_i^{(2)}$	Expression
$H_1^{(1)}$	$C_yD_yT_7$
$H_2^{(1)}$	$2C_yD_yT_6 + Q_yT_4$
$H_3^{(1)}$	$C_yD_yT_5 + 2Q_yT_6 + 3B_yC_yT_7$
$H_4^{(1)}$	$2C_yD_yT_4 + Q_yT_5 + 6(B_yC_y + A_yD_y)T_6 + 2P_yT_7$
$H_5^{(1)}$	$C_yD_yT_3 + 2Q_yT_4 + 3B_yC_yT_5 + 4P_yT_6 + 5A_yB_yT_7$
$H_6^{(1)}$	$2C_yD_yT_2 + Q_yT_3 + 6B_yC_yT_4 + 2P_yT_5 + 10A_yB_yT_6 + 3A_y^2T_7$
$H_7^{(1)}$	$C_yD_yT_1 + 2Q_yT_2 + 3B_yC_yT_3 + 4P_yT_4 + 5A_yB_yT_5 + 6A_y^2T_6$
$H_8^{(1)}$	$Q_yT_1 + 6(B_yC_y + A_yD_y)T_2 + 2P_yT_3 + 10A_yB_yT_4 + 3A_y^2T_5$
$H_9^{(1)}$	$3B_yC_yT_1 + 4P_yT_2 + 5A_yB_yT_3 + 6A_y^2T_4$
$H_{10}^{(1)}$	$2P_yT_1 + 10A_yB_yT_2 + 3A_y^2T_3$
$H_{11}^{(1)}$	$5A_yB_yT_1 + 6A_y^2T_2$
$H_{12}^{(1)}$	$A_y^2T_1$

Table A.6

$H_i^{(3)}$	Expression
$H_1^{(3)}$	$C_xD_yT_7$
$H_2^{(3)}$	$2C_xD_yT_6 + Z_xT_7$
$H_3^{(3)}$	$C_xD_yT_5 + 2Z_xT_6 + W_xT_7$
$H_4^{(3)}$	$2C_xD_yT_4 + Z_xT_5 + 2W_xT_6 + V_xT_7$
$H_5^{(3)}$	$C_xD_yT_3 + 2Z_xT_4 + W_xT_5 + 2V_xT_6 + U_xT_7$
$H_6^{(3)}$	$2C_xD_yT_2 + Z_xT_3 + 2W_xT_4 + V_xT_5 + 2U_xT_6 + 3A_xA_yT_7$
$H_7^{(3)}$	$C_xD_yT_1 + 2Z_xT_2 + W_xT_3 + 2V_xT_4 + U_xT_5 + 6A_xA_yT_6$
$H_8^{(3)}$	$Z_xT_1 + 2W_xT_2 + V_xT_3 + 2U_xT_4 + 3A_xA_yT_5$
$H_9^{(3)}$	$W_xT_1 + 2V_xT_2 + U_xT_3 + 6A_xA_yT_4$
$H_{10}^{(3)}$	$V_xT_1 + 2U_xT_2 + 3A_xA_yT_3$
$H_{11}^{(3)}$	$U_xT_1 + 6A_xA_yT_2$
$H_{12}^{(3)}$	$A_xA_yT_1$

Table A.7

$H_i^{(4)}$	Expression
$H_1^{(4)}$	$C_y D_x T_7$
$H_2^{(4)}$	$2C_y D_x T_6 + Z_y T_7$
$H_3^{(4)}$	$C_y D_x T_5 + 2Z_y T_6 + W_y T_7$
$H_4^{(4)}$	$2C_y D_x T_4 + Z_y T_5 + 2W_y T_6 + V_y T_7$
$H_5^{(4)}$	$C_y D_x T_3 + 2Z_y T_4 + W_y T_5 + 2V_y T_6 + U_y T_7$
$H_6^{(4)}$	$2C_y D_x T_2 + Z_y T_3 + 2W_y T_4 + V_y T_5 + 2U_y T_6 + 3A_y A_x T_7$
$H_7^{(4)}$	$C_y D_x T_1 + 2Z_y T_2 + W_y T_3 + 2V_y T_4 + U_y T_5 + 6A_y A_x T_6$
$H_8^{(4)}$	$Z_y T_1 + 2W_y T_2 + V_y T_3 + 2U_y T_4 + 3A_y A_x T_5$
$H_9^{(4)}$	$W_y T_1 + 2V_y T_2 + U_y T_3 + 6A_y A_x T_4$
$H_{10}^{(4)}$	$V_y T_1 + 2U_y T_2 + 3A_y A_x T_3$
$H_{11}^{(4)}$	$U_y T_1 + 6A_y A_x T_2$
$H_{12}^{(4)}$	$A_y A_x T_1$

Table A.8

Coef_x	Expression
P_x	$B_x^2 + 2A_xC_x$
Q_x	$C_x^2 + 2B_xD_x$
U_x	$2A_yB_x + 3A_xB_y$
V_x	$A_yC_x + 2B_xB_y + 3A_xC_y$
W_x	$B_yC_x + 2B_xC_y$
Z_x	$C_xC_y + 2B_xD_y$

APPENDIX B REVIEW PROBLEMS

1. Introduction

Included in this appendix are some problems aimed at testing how well the subject matter hase been grasped. They range from rather straightforward problems to sophisticated questions requiring a specialized level of knowledge or insight of the underlying issues. The latter are included to arouse research interest on the subject.

2. Review Problems

B.1. Determine the displacement program $s(\psi)$ of the translating roller follower of a cam mechanism with the shape shown in Fig. B.1. In that figure, the lower-dwell and the uniform-velocity programs are to be blended with polynomial P_1, which will produce a smooth transition between the two phases. Moreover, the blending polynomials P_2, P_3, and P_4 are determined similarly. Assume $h = 100$ mm and the data shown in Fig. B.1.

Hint: A large amount of work will be saved if one determines P_2, P_3, and P_4 using the tricks given in Section 2.2, namely translation, scaling, and reflection of P_1.

B.2. Shown in Fig. B.2 is a portion of a symmetric displacement program for the translating follower of a cam mechanism. It is called symmetric because its return phase is the mirror image of the rise phase with respect to the centerline shown. Find the polynomial motions $P_1(\psi)$ and $P_2(\psi)$ that will produce a smooth blending between lines L_1 and L_2 and the upper dwell D_u, which occurs for $s(\psi) = h$.
Hint: Exploit the symmetries of the problem in order to reduce your computational work to a minimum.

B.3. Throughout the book it was made apparent that 3-4-5 polynomial motions yield smaller minimum-size cams than their cyloidal counterparts. Investigate if other motion programs exist, with the same smooth properties of the two aforementioned programs at the end points of either the rise or the descent phase, yielding even smaller minimum-size cams. Answering this question may require resorting to calculus of variations (Gelfand and Fomin, 1963) or optimal control techniques (Bryson and Ho, 1975).

B.4. Shown in Fig. B.3 is a cam mechanism with a translating roller follower. The cam profile is a circle with center at C which rotates about point O.

B.4.1. Find an expression for the pressure angle as a function of ψ and the mechanism parameters, a, b, c, and e.

B.4.2. Find the maximum and minimum values of the pressure angle and the values of ψ at which these extrema are attained.

B.4.3. Find the numerical value of the maximum pressure angle using the following numerical values: $a = 10$ mm, $b = 40$ mm, $c = 20$ mm, and $e = 30$ mm.

B.4.4. If the pressure angle should not exceed 30°, is the foregoing design acceptable? If not, redesign the cam mechanism by properly redefining a, b, c, and e. If this is done, will the displacement program of the follower be changed?

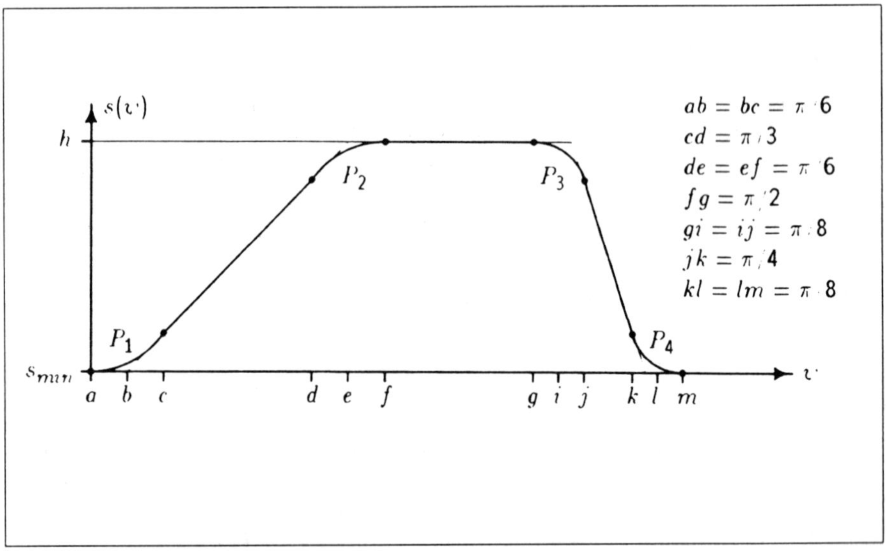

Fig. B.1. Displacement program of a translating follower.

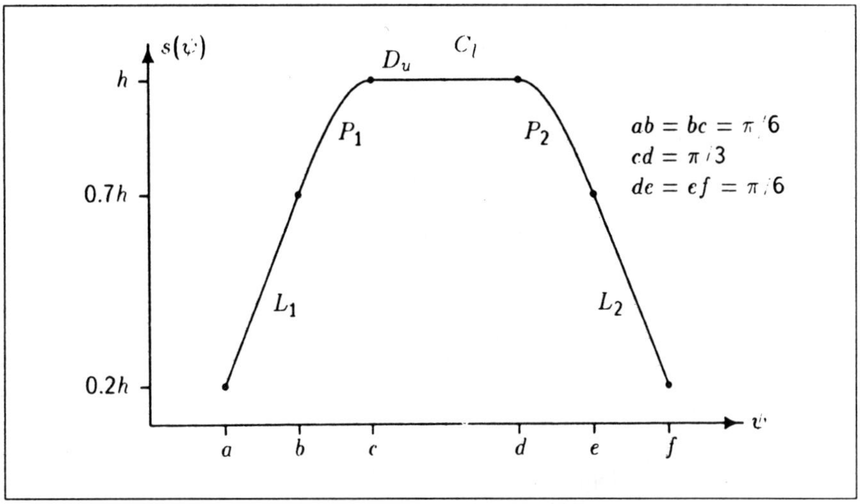

Fig. B.2. Symmetric displacement program.

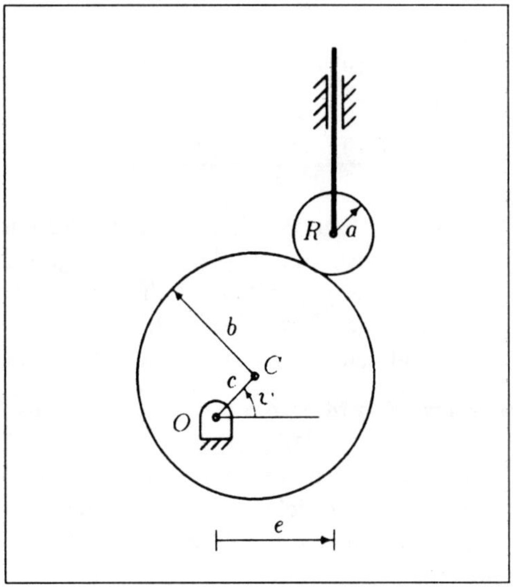

Fig. B.3. Translating roller-follower cam mechanism.

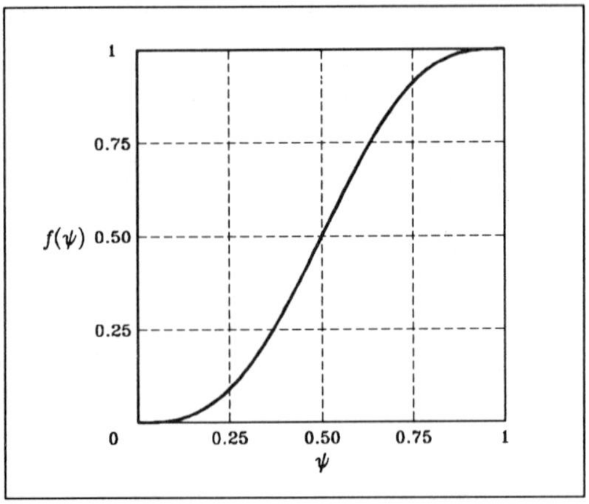

Fig. B.4. Function for Problem B.5.

B.5. Given in Fig. B.4 is a smooth function of x, $f(x)$, with the following properties:

$$f(0) = 0, \qquad f(1) = 1$$

$$f'(0) = 0, \qquad f'(1) = 0$$

$$f''(0) = 0, \qquad f''(1) = 0$$

Write a procedure that admits function $f(\cdot)$ as input, along with a value x of its argument, $0 \leq x \leq 1$, and returns the displacement program, $\sigma(\psi)$, beginning at $\psi = 0$ and ending at $\psi = 2\pi$, of the motion of the follower of a cam mechanism, with the given motion percentages: lower dwell, D_1 % (dwell begins at $\psi = 0$); rise, R_1 %; upper dwell, D_2 %; return, R_2 % (return ends at $\psi = 2\pi$), for a rise h.

Hint: Use the tricks described in Section 2.2.

B.6. Write a code to produce the follower displacement program of an indexing cam mechanism with an oscillating roller-follower. To this effect, use a 3-4-5 polynomial motion, with N indexing steps and a duration of the dwells equal to one-half the duration of the rise phases.

B.7. Now assume that the follower of Problem B.6 undergoes $2N$ indexing steps, but reverses its velocity after N steps. Find the displacement program of this motion under the assumption that it is symmetric.

B.8. Design the rise motion program of the follower of a cam mechanism based on the shape taken on by a clamped-clamped beam under a uniformly distributed load.

B.9. Investigate the applicability of *trigonometric polynomials* $T_n(x)$ of degree n, defined below, to the synthesis of the follower motion program.

$$T_n(x) \equiv \sum_0^n a_k \cos 2\pi kx + \sum_1^n b_k \sin 2\pi kx$$

Try to answer the questions:

i) Are these polynomials better conditioned than ordinary polynomials, for our purposes?;

ii) Are they better conditioned than orthogonal polynomials?;

iii) Are trigonometric polynomials orthogonal?

B.10. Sketch a plot of $\cos\alpha$ vs. ψ, where α denotes the pressure angle of a cam mechanism with a translating knife-edge follower undergoing a constant-velocity displacement program and ψ is the angle of rotation of the cam disk.

B.11. Find the value(s) of the normalized offset e/h and the normalized parameter c/h of a cam mechanism with a knife-edge translating follower that will produce a maximum value of the pressure angle of 30°, if the displacement program is harmonic, for arbitrary values of the rise, h.

B.12. Shown in Fig. B.5 is a cam mechanism with an oscillating roller-follower. If the cam disk is an offset circular plate with its center at C, find an expression for the pressure angle, α, in terms of ψ and the mechanism parameters.

B.13. Use the procedure you wrote in **B.5**, to produce the displacement programs given in the examples of Subsections 4.2.4, 4.3.3, 5.2.5, and 5.3.1.

B.14. Write a code to generate the cam profile and the *local* geometric properties of the example given in Subsection 4.2.4.

B.15. Idem Problem B.14 but for the data given in Subsection 4.3.3. You will have to write a code to obtain the cam profile for the roller-follower and compute the maximum curvature and the tangent and normal unit vectors.

B.16. Idem Problem B.14 but for the data given in Subsection 5.2.5.

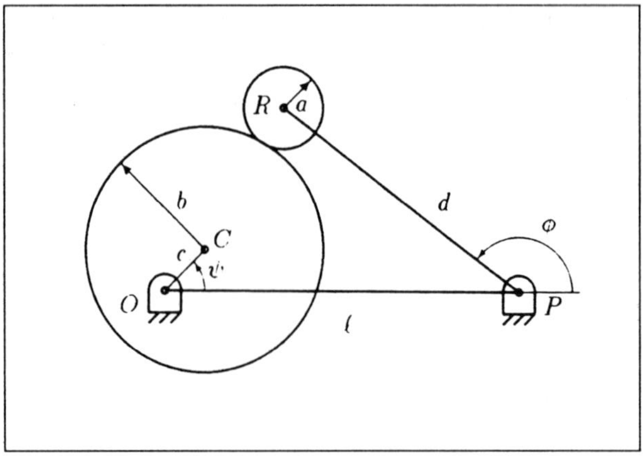

Fig. B.5. Cam mechanism with an oscillating roller follower.

B.17. Idem Problem B.14 but for the data given in Subsection 5.3.1. You will have to develop a code for obtaining the optimum parameters for this type of cam mechanism.

B.18. Idem Problem B.14 but with the data given in Problem B.4.

B.19. Give a definition for ϵ of Fig. 1.4.3*b*. List the reasons why this variable should be taken into account. In addition, outline a method of computing ϵ in terms of the mechanism parameters.

B.20. Determine the profile of a cam with a translating roller follower undergoing a rise-dwell-return-dwell motion with the following percentages: rise, 30%; dwell, 30%; return, 20%; dwell, 20%. Moreover, the rise and the return phases are cycloidal motions, and the pressure angle should attain a maximum of $+45°$ and a minimum of $-45°$.

B.21. The cam-operated feeding mechanism of a machine tool should perform a machining operation by driving the tool into the workpiece to be machined following a preestablished motion program. Once the machining is completed, the tool should return to its original position in order to allow for a new workpiece to be fed into the machine. In order to complete this workpiece-feeding operation, the tool should remain at rest in its original position, in order to allow for the workpiece to be properly fastened to the machine frame, thereby completing one cycle. Moreover, this program, to be described below, is to be repeated periodically, which is the reason why a cam-operated mechanism was chosen in the first place. Each cycle consists of three stages, namely,

1. Rise phase. This takes place in 50% of the total cycle, its purpose being to cut metal from the workpiece. This is the working phase of the total cycle, and hence, in this phase, the tool is under full load. Obviously, in order to avoid shock and collisions, the tool should start and end its motion with zero velocity and acceleration. Moreover, the cutting depth is 20 mm.

2. Return Phase. Since this is the no-load phase of the cycle, the tool can move faster than it did in the rise phase, and hence, a duration of 20% of this phase is assumed.

3. Lower Dwell. The purpose of this phase, which has a duration of the remaining 30% of the cycle, is to allow the feeding of a new workpiece into the machine.

Synthesize the cam mechanism that will produce the foregoing motion program, using a roller-follower, which is the most suitable type for the high loads that will be transmitted to the mechanism. Moreover, you should choose the offset and the radius of the base circle that *i)* will produce the foregoing motion; *ii)* will do so with a pressure angle not exceeding an absolute value of 30° and *iii)* will yield a cam profile of minimum size.

Note that no particular motion program is specified, and there are many that could do the job. Try at least two and compare them in terms of *i)* cam size; *ii)* maximum acceleration; and *iii)* maximum curvature. For each design that you propose, specify the maximum allowable radius of the mill with which the cam profile is to be cut. Computer-generated plots of the items listed below are required:

- The motion programs of your choice;
- The cam profiles of minimum size;
- The trajectory of the cutting mill for each profile; and
- The distribution of the pressure angle throughout the cam profile.

B.22. Design of a positive-motion cam with a translating follower.
A cam mechanism with a translating roller-follower is shown in Fig. 1.3.5*b*. During the rise phase, contact is guaranteed by the opposite directions of the contact force exerted by the cam on the follower and the load, which is assumed to act downwards. However, at the return phase, the two forces no longer oppose each other, and contact can be lost. To avoid loss of contact, a spring can be used, but springs introduce low natural frequencies in the mechanism, and hence, should be avoided. An alternative to springs are positive-motion cams, whereby the roller moves in a track grooved in the cam plate. This solution, nevertheless, brings about undesirable friction forces and should also be avoided.

Investigate the use of a *protruded* track in the cam plate, similar to that of Fig. 6.2.9 for a cylindrical cam and produce drawings and, if possible, a machined cam plate, of your design, for the specifications in Example 5.2.3.

B.23. Idem Problem B.22 for a cam with an oscillating roller-follower. Use the specifications given in Example 5.3.2.

B.24. Note that, in the synthesis equations for the cam with an oscillating flat-face follower, the numerator and the denominator of both $\rho(\theta)$ and $\tan(\psi+\phi+\theta)$ vanish if $\phi'(\psi)+1$ vanishes, which in turn happens when $\psi = 90°$. Derive expressions for these variables that will allow you to calculate the polar and the Cartesian coordinates of the cam profile under the said conditions.

Hint: Use L'Hospital's rule.

B.25. A bound for $\overline{e}$ was derived in (Angeles et al. 1989b) that nevertheless is too conservative. This bound was based on a discriminant associated with the quadratic equation arising from the zeroing of function $F(\psi)$ given in eq.(4.3.31*b*), upon substitution of $\cos\phi$ and $\sin\phi$ by their corresponding expressions in terms of $\tan(\phi/2)$. Note, however, that the coefficients of the quadratic equation are implicitly functions of ϕ, for they contain $\phi'(\psi)$ and $\phi''(\psi)$. Alternatively, a quadratic equation in $T \equiv \tan(\beta/2)$ that is independent of ψ can be derived if ϕ is substituted by $\beta + \sigma$, as given by eq.(4.3.1). Show that the quadratic equation shown below is derived:

$$A(\psi)\,T^2 + 2\,B(\psi)\,T + C(\psi) = 0$$

whose coefficients are given by

$$A(\psi) \equiv \overline{e}\,(1+\phi')^3 - \phi''\cos\sigma - (1+\phi')\,(1+2\phi')\sin\sigma$$

$$B(\psi) \equiv -\phi''\sin\sigma + (1+\phi')\,(1+2\phi')\sin\sigma$$

$$A(\psi) \equiv \overline{e}\,(1+\phi')^3 + \phi''\cos\sigma + (1+\phi')\,(1+2\,\phi')\sin\sigma$$

In addition, show that the same bound for $\overline{e}$ is found upon imposing the condition that the discriminant of the foregoing equation vanishes, i.e.,

$$\overline{e} < \sqrt{\max_{\psi} G(\psi)}$$

with $G(\psi)$ defined as

$$G(\psi) \equiv \frac{[(1+\phi')\,(1+2\phi')\,]^2 + \phi''^2}{(1+\phi')^6}$$

INDEX

Mechanics

SOLID MECHANICS AND ITS APPLICATIONS

Series Editor: G.M.L. Gladwell

Aims and Scope of the Series

The fundamental questions arising in mechanics are: *Why?, How?,* and *How much?* The aim of this series is to provide lucid accounts written by authoritative researchers giving vision and insight in answering these questions on the subject of mechanics as it relates to solids. The scope of the series covers the entire spectrum of solid mechanics. Thus it includes the foundation of mechanics; variational formulations; computational mechanics; statics, kinematics and dynamics of rigid and elastic bodies; vibrations of solids and structures; dynamical systems and chaos; the theories of elasticity, plasticity and viscoelasticity; composite materials; rods, beams, shells and membranes; structural control and stability; soils, rocks and geomechanics; fracture; tribology; experimental mechanics; biomechanics and machine design.

1. R.T. Haftka, Z. Gürdal and M.P. Kamat: *Elements of Structural Optimization.* 2nd rev.ed., 1990 ISBN 0-7923-0608-2
2. J.J. Kalker: *Three-Dimensional Elastic Bodies in Rolling Contact.* 1990 ISBN 0-7923-0712-7
3. P. Karasudhi: *Foundations of Solid Mechanics.* 1991 ISBN 0-7923-0772-0
4. N. Kikuchi: *Computational Methods in Contact Mechanics.* (forthcoming) ISBN 0-7923-0773-9
5. Y.K. Cheung and A.Y.T. Leung: *Finite Element Methods in Dynamics.* (forthcoming) ISBN 0-7923-1313-5
6. J.F. Doyle: *Static and Dynamic Analysis of Structures.* With an Emphasis on Mechanics and Computer Matrix Methods. 1991 ISBN 0-7923-1124-8; Pb 0-7923-1208-2
7. O.O. Ochoa and J.N. Reddy: *Finite Element Modelling of Composite Structures.* (forthcoming) ISBN 0-7923-1125-6
8. M.H. Aliabadi and D.P. Rooke: *Numerical Fracture Mechanics.* ISBN 0-7923-1175-2
9. J. Angeles and C.S. López-Cajún: *Optimization of Cam Mechanisms.* 1991 ISBN 0-7923-1355-0
10. D.E. Grierson, A. Franchi and P. Riva: *Progress in Structural Engineering.* 1991 ISBN 0-7923-1396-8

Kluwer Academic Publishers – Dordrecht / Boston / London

Mechanics

FLUID MECHANICS AND ITS APPLICATIONS

Series Editor: R. Moreau

Aims and Scope of the Series

The purpose of this series is to focus on subjects in which fluid mechanics plays a fundamental role. As well as the more traditional applications of aeronautics, hydraulics, heat and mass transfer etc., books will be published dealing with topics which are currently in a state of rapid development, such as turbulence, suspensions and multiphase fluids, super and hypersonic flows and numerical modelling techniques. It is a widely held view that it is the interdisciplinary subjects that will receive intense scientific attention, bringing them to the forefront of technological advancement. Fluids have the ability to transport matter and its properties as well as transmit force, therefore fluid mechanics is a subject that is particularly open to cross fertilisation with other sciences and disciplines of engineering. The subject of fluid mechanics will be highly relevant in domains such as chemical, metallurgical, biological and ecological engineering. This series is particularly open to such new multidisciplinary domains.

1. M. Lesieur: *Turbulence in Fluids.* 2nd rev. ed., 1990 — ISBN 0-7923-0645-7
2. O. Métais and M. Lesieur (eds.): *Turbulence and Coherent Structures.* 1991 — ISBN 0-7923-0646-5
3. R. Moreau: *Magnetohydrodynamics.* 1990 — ISBN 0-7923-0937-5
4. E. Coustols (ed.): *Turbulence Control by Passive Means.* 1990 — ISBN 0-7923-1020-9
5. A. A. Borissov (ed.): *Dynamic Structure of Detonation in Gaseous and Dispersed Media.* 1991 — ISBN 0-7923-1340-2

Kluwer Academic Publishers – Dordrecht / Boston / London

Mechanics

From 1990, books on the subject of *mechanics* will be published under two series:
***FLUID* MECHANICS AND ITS APPLICATIONS**
Series Editor: R.J. Moreau
***SOLID* MECHANICS AND ITS APPLICATIONS**
Series Editor: G.M.L. Gladwell

Prior to 1990, the books listed below were published in the respective series indicated below.

MECHANICS: DYNAMICAL SYSTEMS
Editors: L. Meirovitch and G.Æ. Oravas

1. E.H. Dowell: *Aeroelasticity of Plates and Shells.* 1975 ISBN 90-286-0404-9
2. D.G.B. Edelen: *Lagrangian Mechanics of Nonconservative Nonholonomic Systems.* 1977 ISBN 90-286-0077-9
3. J.L. Junkins: *An Introduction to Optimal Estimation of Dynamical Systems.* 1978 ISBN 90-286-0067-1
4. E.H. Dowell (ed.), H.C. Curtiss Jr., R.H. Scanlan and F. Sisto: *A Modern Course in Aeroelasticity.* *Revised and enlarged edition see under Volume 11*
5. L. Meirovitch: *Computational Methods in Structural Dynamics.* 1980 ISBN 90-286-0580-0
6. B. Skalmierski and A. Tylikowski: *Stochastic Processes in Dynamics.* Revised and enlarged translation. 1982 ISBN 90-247-2686-7
7. P.C. Müller and W.O. Schiehlen: *Linear Vibrations.* A Theoretical Treatment of Multi-degree-of-freedom Vibrating Systems. 1985 ISBN 90-247-2983-1
8. Gh. Buzdugan, E. Mihăilescu and M. Radeş: *Vibration Measurement.* 1986 ISBN 90-247-3111-9
9. G.M.L. Gladwell: *Inverse Problems in Vibration.* 1987 ISBN 90-247-3408-8
10. G.I. Schuëller and M. Shinozuka: *Stochastic Methods in Structural Dynamics.* 1987 ISBN 90-247-3611-0
11. E.H. Dowell (ed.), H.C. Curtiss Jr., R.H. Scanlan and F. Sisto: *A Modern Course in Aeroelasticity.* Second revised and enlarged edition (of Volume 4). 1989 ISBN Hb 0-7923-0062-9; Pb 0-7923-0185-4
12. W. Szemplińska-Stupnicka: *The Behavior of Nonlinear Vibrating Systems.* Volume I: Fundamental Concepts and Methods: Applications to Single-Degree-of-Freedom Systems. 1990 ISBN 0-7923-0368-7
13. W. Szemplińska-Stupnicka: *The Behavior of Nonlinear Vibrating Systems.* Volume II: Advanced Concepts and Applications to Multi-Degree-of-Freedom Systems. 1990 ISBN 0-7923-0369-5
Set ISBN (Vols. 12–13) 0-7923-0370-9

MECHANICS OF STRUCTURAL SYSTEMS
Editors: J.S. Przemieniecki and G.Æ. Oravas

1. L. Frýba: *Vibration of Solids and Structures under Moving Loads.* 1970 ISBN 90-01-32420-2
2. K. Marguerre and K. Wölfel: *Mechanics of Vibration.* 1979 ISBN 90-286-0086-8

Mechanics

3. E.B. Magrab: *Vibrations of Elastic Structural Members.* 1979 ISBN 90-286-0207-0
4. R.T. Haftka and M.P. Kamat: *Elements of Structural Optimization.* 1985
   Revised and enlarged edition see under Solid Mechanics and Its Applications, Volume 1
5. J.R. Vinson and R.L. Sierakowski: *The Behavior of Structures Composed of Composite Materials.* 1986 ISBN Hb 90-247-3125-9; Pb 90-247-3578-5
6. B.E. Gatewood: *Virtual Principles in Aircraft Structures.* Volume 1: Analysis. 1989
   ISBN 90-247-3754-0
7. B.E. Gatewood: *Virtual Principles in Aircraft Structures.* Volume 2: Design, Plates, Finite Elements. 1989 ISBN 90-247-3755-9
   Set (Gatewood 1 + 2) ISBN 90-247-3753-2

MECHANICS OF ELASTIC AND INELASTIC SOLIDS
Editors: S. Nemat-Nasser and G.Æ. Oravas

1. G.M.L. Gladwell: *Contact Problems in the Classical Theory of Elasticity.* 1980
   ISBN Hb 90-286-0440-5; Pb 90-286-0760-9
2. G. Wempner: *Mechanics of Solids with Applications to Thin Bodies.* 1981
   ISBN 90-286-0880-X
3. T. Mura: *Micromechanics of Defects in Solids.* 2nd revised edition, 1987
   ISBN 90-247-3343-X
4. R.G. Payton: *Elastic Wave Propagation in Transversely Isotropic Media.* 1983
   ISBN 90-247-2843-6
5. S. Nemat-Nasser, H. Abé and S. Hirakawa (eds.): *Hydraulic Fracturing and Geothermal Energy.* 1983 ISBN 90-247-2855-X
6. S. Nemat-Nasser, R.J. Asaro and G.A. Hegemier (eds.): *Theoretical Foundation for Large-scale Computations of Nonlinear Material Behavior.* 1984 ISBN 90-247-3092-9
7. N. Cristescu: *Rock Rheology.* 1988 ISBN 90-247-3660-9
8. G.I.N. Rozvany: *Structural Design via Optimality Criteria.* The Prager Approach to Structural Optimization. 1989 ISBN 90-247-3613-7

MECHANICS OF SURFACE STRUCTURES
Editors: W.A. Nash and G.Æ. Oravas

1. P. Seide: *Small Elastic Deformations of Thin Shells.* 1975 ISBN 90-286-0064-7
2. V. Panc: *Theories of Elastic Plates.* 1975 ISBN 90-286-0104-X
3. J.L. Nowinski: *Theory of Thermoelasticity with Applications.* 1978
   ISBN 90-286-0457-X
4. S. Łukasiewicz: *Local Loads in Plates and Shells.* 1979 ISBN 90-286-0047-7
5. C. Fiřt: *Statics, Formfinding and Dynamics of Air-supported Membrane Structures.* 1983 ISBN 90-247-2672-7
6. Y. Kai-yuan (ed.): *Progress in Applied Mechanics.* The Chien Wei-zang Anniversary Volume. 1987 ISBN 90-247-3249-2
7. R. Negruţiu: *Elastic Analysis of Slab Structures.* 1987 ISBN 90-247-3367-7
8. J.R. Vinson: *The Behavior of Thin Walled Structures.* Beams, Plates, and Shells. 1988
   ISBN Hb 90-247-3663-3; Pb 90-247-3664-1

Mechanics

MECHANICS OF FLUIDS AND TRANSPORT PROCESSES
Editors: R.J. Moreau and G.Æ. Oravas

1. J. Happel and H. Brenner: *Low Reynolds Number Hydrodynamics*. With Special Applications to Particular Media. 1983 ISBN Hb 90-01-37115-9; Pb 90-247-2877-0
2. S. Zahorski: *Mechanics of Viscoelastic Fluids*. 1982 ISBN 90-247-2687-5
3. J.A. Sparenberg: *Elements of Hydrodynamics Propulsion*. 1984 ISBN 90-247-2871-1
4. B.K. Shivamoggi: *Theoretical Fluid Dynamics*. 1984 ISBN 90-247-2999-8
5. R. Timman, A.J. Hermans and G.C. Hsiao: *Water Waves and Ship Hydrodynamics*. An Introduction. 1985 ISBN 90-247-3218-2
6. M. Lesieur: *Turbulence in Fluids*. Stochastic and Numerical Modelling. 1987 ISBN 90-247-3470-3
7. L.A. Lliboutry: *Very Slow Flows of Solids*. Basics of Modeling in Geodynamics and Glaciology. 1987 ISBN 90-247-3482-7
8. B.K. Shivamoggi: *Introduction to Nonlinear Fluid-Plasma Waves*. 1988 ISBN 90-247-3662-5
9. V. Bojarevičs, Ya. Freibergs, E.I. Shilova and E.V. Shcherbinin: *Electrically Induced Vortical Flows*. 1989 ISBN 90-247-3712-5
10. J. Lielpeteris and R. Moreau (eds.): *Liquid Metal Magnetohydrodynamics*. 1989 ISBN 0-7923-0344-X

MECHANICS OF ELASTIC STABILITY
Editors: H. Leipholz and G.Æ. Oravas

1. H. Leipholz: *Theory of Elasticity*. 1974 ISBN 90-286-0193-7
2. L. Librescu: *Elastostatics and Kinetics of Aniosotropic and Heterogeneous Shell-type Structures*. 1975 ISBN 90-286-0035-3
3. C.L. Dym: *Stability Theory and Its Applications to Structural Mechanics*. 1974 ISBN 90-286-0094-9
4. K. Huseyin: *Nonlinear Theory of Elastic Stability*. 1975 ISBN 90-286-0344-1
5. H. Leipholz: *Direct Variational Methods and Eigenvalue Problems in Engineering*. 1977 ISBN 90-286-0106-6
6. K. Huseyin: *Vibrations and Stability of Multiple Parameter Systems*. 1978 ISBN 90-286-0136-8
7. H. Leipholz: *Stability of Elastic Systems*. 1980 ISBN 90-286-0050-7
8. V.V. Bolotin: *Random Vibrations of Elastic Systems*. 1984 ISBN 90-247-2981-5
9. D. Bushnell: *Computerized Buckling Analysis of Shells*. 1985 ISBN 90-247-3099-6
10. L.M. Kachanov: *Introduction to Continuum Damage Mechanics*. 1986 ISBN 90-247-3319-7
11. H.H.E. Leipholz and M. Abdel-Rohman: *Control of Structures*. 1986 ISBN 90-247-3321-9
12. H.E. Lindberg and A.L. Florence: *Dynamic Pulse Buckling*. Theory and Experiment. 1987 ISBN 90-247-3566-1
13. A. Gajewski and M. Zyczkowski: *Optimal Structural Design under Stability Constraints*. 1988 ISBN 90-247-3612-9

Mechanics

MECHANICS: ANALYSIS

Editors: V.J. Mizel and G.Æ. Oravas

1. M.A. Krasnoselskii, P.P. Zabreiko, E.I. Pustylnik and P.E. Sbolevskii: *Integral Operators in Spaces of Summable Functions*. 1976 ISBN 90-286-0294-1
2. V.V. Ivanov: *The Theory of Approximate Methods and Their Application to the Numerical Solution of Singular Integral Equations*. 1976 ISBN 90-286-0036-1
3. A. Kufner, O. John and S. Pučík: *Function Spaces*. 1977 ISBN 90-286-0015-9
4. S.G. Mikhlin: *Approximation on a Rectangular Grid*. With Application to Finite Element Methods and Other Problems. 1979 ISBN 90-286-0008-6
5. D.G.B. Edelen: *Isovector Methods for Equations of Balance*. With Programs for Computer Assistance in Operator Calculations and an Exposition of Practical Topics of the Exterior Calculus. 1980 ISBN 90-286-0420-0
6. R.S. Anderssen, F.R. de Hoog and M.A. Lukas (eds.): *The Application and Numerical Solution of Integral Equations*. 1980 ISBN 90-286-0450-2
7. R.Z. Has'minskiĭ: *Stochastic Stability of Differential Equations*. 1980 ISBN 90-286-0100-7
8. A.I. Vol'pert and S.I. Hudjaev: *Analysis in Classes of Discontinuous Functions and Equations of Mathematical Physics*. 1985 ISBN 90-247-3109-7
9. A. Georgescu: *Hydrodynamic Stability Theory*. 1985 ISBN 90-247-3120-8
10. W. Noll: *Finite-dimensional Spaces*. Algebra, Geometry and Analysis. Volume I. 1987 ISBN Hb 90-247-3581-5; Pb 90-247-3582-3

MECHANICS: COMPUTATIONAL MECHANICS

Editors: M. Stern and G.Æ. Oravas

1. T.A. Cruse: *Boundary Element Analysis in Computational Fracture Mechanics*. 1988 ISBN 90-247-3614-5

MECHANICS: GENESIS AND METHOD

Editor: G.Æ. Oravas

1. P.-M.-M. Duhem: *The Evolution of Mechanics*. 1980 ISBN 90-286-0688-2

MECHANICS OF CONTINUA

Editors: W.O. Williams and G.Æ. Oravas

1. C.-C. Wang and C. Truesdell: *Introduction to Rational Elasticity*. 1973 ISBN 90-01-93710-1
2. P.J. Chen: *Selected Topics in Wave Propagation*. 1976 ISBN 90-286-0515-0
3. P. Villaggio: *Qualitative Methods in Elasticity*. 1977 ISBN 90-286-0007-8

Mechanics

MECHANICS OF FRACTURE

Editors: G.C. Sih

1. G.C. Sih (ed.): *Methods of Analysis and Solutions of Crack Problems.* 1973 ISBN 90-01-79860-8
2. M.K. Kassir and G.C. Sih (eds.): *Three-dimensional Crack Problems.* A New Solution of Crack Solutions in Three-dimensional Elasticity. 1975 ISBN 90-286-0414-6
3. G.C. Sih (ed.): *Plates and Shells with Cracks.* 1977 ISBN 90-286-0146-5
4. G.C. Sih (ed.): *Elastodynamic Crack Problems.* 1977 ISBN 90-286-0156-2
5. G.C. Sih (ed.): *Stress Analysis of Notch Problems.* Stress Solutions to a Variety of Notch Geometries used in Engineering Design. 1978 ISBN 90-286-0166-X
6. G.C. Sih and E.P. Chen (eds.): *Cracks in Composite Materials.* A Compilation of Stress Solutions for Composite System with Cracks. 1981 ISBN 90-247-2559-3
7. G.C. Sih (ed.): *Experimental Evaluation of Stress Concentration and Intensity Factors.* Useful Methods and Solutions to Experimentalists in Fracture Mechanics. 1981 ISBN 90-247-2558-5

MECHANICS OF PLASTIC SOLIDS

Editors: J. Schroeder and G.Æ. Oravas

1. A. Sawczuk (ed.): *Foundations of Plasticity.* 1973 ISBN 90-01-77570-5
2. A. Sawczuk (ed.): *Problems of Plasticity.* 1974 ISBN 90-286-0233-X
3. W. Szczepiński: *Introduction to the Mechanics of Plastic Forming of Metals.* 1979 ISBN 90-286-0126-0
4. D.A. Gokhfeld and O.F. Cherniavsky: *Limit Analysis of Structures at Thermal Cycling.* 1980 ISBN 90-286-0455-3
5. N. Cristescu and I. Suliciu: *Viscoplasticity.* 1982 ISBN 90-247-2777-4

Kluwer Academic Publishers - Dordrecht / Boston / London